Molluscan Diversity of Chilika Lagoon

Molluscan Diversity of Chilika Lagoon

Dr. Debasish Mahapatra
Dr. Surendra S. Kadam

AAYU PUBLICATIONS
New Delhi

Cataloging in Publication Data--DK

Courtesy: D.K. Agencies (P) Ltd. <docinfo@dkagencies.com>

Mahapatra, Debasish, *Dr.*, author.
Molluscan diversity of Chilika lagoon / authors, Dr. Debasish Mahapatra, Dr. Surendra S. Kadam. -- 1st edition.
pages cm
Includes bibliographical references.
ISBN 9789385161872

1. Mollusks--India--Chilka Lake. I. Kadam, Surendra S., author. II. Title.

LCC QL403.M34 2018 | DDC 594.0954 23

Published by:

AAYU PUBLICATIONS
D-134, Agar Nagar, Prem Nagar-III
New Delhi-110086
Tel: 09910728725, 09654504227
Email: aayupublication@gmail.com
www.aayupublications.com

Molluscan Diversity of Chilika Lagoon

1st Edition 2018

ISBN: 978-93-85161-87-2

PRINTED IN INDIA

Printed at: Unitech Graphic Point, Delhi

महाराष्ट्र विधानसभा

विधान सभा
महाराष्ट्र

श्री. अमित कृष्णा घोडा
१३०, (अ.ज.) पालघर
आमदार

मु.पो. रानशेत, ता. डहाणू, जि. पालघर - ४०१ ६०२. मो.: ९७६५४४५५३२ / ९२६०४४५५३२
E-mail : amitghoda1982@gmail.com

जा. क्र. : २८७४/२०१८

दिनांक : २६/१०/२०१८

MESSAGE

Chilika lagoon is the largest brackish water ecosystem in Asia popular for migratory birds , Irrawaddy dolphins, fish diversity and abundance. Molluscan diversity of Chilika is also very rich but has not been addressed properly. Mollusca are become the integral part of our life since time immemorial. They are used as lime in beetle leaf and white washing of the houses. Sacred chank is used as religious purpose while cowries are used as a form of alternative source of transactions for trading in ancient India. I am happy that the book entitled "Molluscan Diversity of Chilika Lagoon" authored by Dr. Debasish Mahapatro and Dr. Surendra S. Kadam contains all the detailed information regarding mollusca which will be helpful for the school and college students, researchers, professors, environmentalists etc for their academic and research purposes. They have given a number of colourful photographs of mollusc along with their classification, description and distribution at the present scenario in a very concise manner. This book is attractive as well as informative.

I wish the publication all success.

Signature

Shri. Amit Krushna Ghoda

(M. L. A. Palghar)

Contents

Foreword

When my colleague, Dr. Surendra S. Kadam showed me the book entitled "Molluscan Diversity of Chilika Lagoon" co-authored by Dr. Debasish Mahapatro, I became more enthusiastic to know the molluscan community of a coastal lagoon like Chilika. The layout and design, colorful illustrations of different species of mollusca is well furnished in the book. It also contains a detailed taxonomic classification, description and the distribution of species in India and other parts of the world. I strongly feel that the embedded with concise and comprehensive information on mollusc which will be very essential to the students from school level to undergraduate and post graduate, scientists, academicians, environmentalists and even to the policy makers. This book can be used as a pictorial guide to the molluscan lovers even for the common man. Although, my acquaintance with mollusc is very limited to the shells of cowries which was used as a form of money in the trade and commerce of ancient India. However, after reading this book, I am much encouraged to know more about these shelled organisms. Mollusc plays a vital role in the ecological functioning as well as food for human consumption, as a piece of decoration, also acts as an indicator of environmental pollution etc. Molluscan fishery further adds revenue to the country and its aquaculture practice provides an alternative livelihood for the fishermen hence contributing to the socio-economic development of the fishing community.

Being a professor in English, I am remembering one poem "of Molluscs" penned by May Satrons (1912-1995) i.e.

"As the tide rises, the closed mollusc
Opens a fraction to the ocean's food,
Bathed in its riches. Do not ask
What force would do, or if force could".....

I wish this book will get all success.

Dr. Mrs. Anjali Kulkarni
Principal G.E. Society's
N.B. Meheta Science College, Bordi,
Maharastra

Place: **Maharastra**
Date: 22nd January 2018

Preface

When I first visited with my father and mother to Jagannatha Dham, Puri located at the close vicinity of Bay of Bengal in the state of Odisha, I come across the first touch of the sea water and the dead molluscan shells scattered in the beach. Unknowingly, I have collected some those of shells although I was just 7 years of old. For many years after schooling those shells are unused materials for me. But, as the time progresses my interest towards sea and sea shell got augmented. However, real interest to study the sea and its life began during my college days particularly while perusing my graduation in Gopalpur College, Gopalpur-on-Sea. This is located at the Ganjam district which is the southern province of Odisha state, globally well known as the land fall point of the super cyclone *Phailin* stroked on dated 12th of October 2013. The dwindling waves of the sea, the sandy expanses, those beautiful fishes; crabs and prawns all had fascinated me to make my carrier as a marine biologist. When I joined in the P.G. Department of Marine Science, Berhampur University, Odisha for post graduation I got an opportunity to pursue my M.Phil and Ph.D. work. My interest on mollusc augmented further. This is my Ph.D. pursuing period where I feel that it is a great opportunity to fulfil my childhood passion for collecting the sea shells. But, simultaneously get discouraged when understood that there is big gap of information related to molluscan identification. Then I started collection of shells of bivalves and gastropods from the Chilika lagoon during various sampling programs and then tried to identify them by my own interest. And finally I got something different, something new to Chilika. These species are *Aplysia daciylomela*, *Armina maculata*, *Stellaria solaris* and few of the Cowries. These are uncommon in the north east coast of Odisha in general and Chilika lagoon in particular. Chilika faced a sever environmental degradation during the last part of the 20th century because of the lowering the salinity profile. This was occurred due to the ceasing of the sea-lagoon interaction. But, later towards the beginning of the 21st century the lagoon got a vibrant life after opening of an artificial inlet during 23rd of September 2000 and a natural inlet on 1st of August 2008 at the one km northward from the artificially dredged inlet. This is the time for the finding of some amazed species of mollusc from Chilika. But, again problem prevails in the identification of species. Somehow the problem is solved leaving a large bulk of knowledge providing experiences. With the heavy compression of memories from the childhood to till date, I have tried to summarise the entire experience and then converted it in the form of a book entitled "Molluscan Diversity of Chilika Lagoon" authored by me along with Dr. Surendra S. Kadam who is presently working as an Assistant Professor in west coast of India. He is a replica of mine with similar passion and fascination. We have tried to provide good quality pictures of the bivalves and gastropods, their classification, identification, distribution details and all those information which were gathered during my Ph.D program i.e. from 2007 to 2010 under the guidance of Dr. Sudarsan Panda, IFS and Prof. (Retd.) R.C. Panigrahy. I and Dr. Kadam have designed this for wide range readers such as for a layman it will be helpful as a pictorial guide; a popular science book for the students; text book for the undergraduate and postgraduate students; research documentation for the researchers and a valuable document for the scientists. Since, there is a large scale of morphological diversities exists between the different families of bivalves and gastropods and proper identification of all these species becomes a menace for the researchers, students and even to a common man. Therefore, the present work will help to fill the missing link.

This book is the reality of a dream viewed by a child while moving in a beach and now as an author begs to the almighty that it may get a splendid success.

Dr. Debasish Mahapatra

Dr. Surendra S. Kadam

Acknowledgement

This book is the end of my journey; I have not travelled in a vacuum in this journey as everywhere during the journey. I came across the people who acted as mentor & inspiring force during writing of this book.

First of all, I express my deep sense of gratitude to my research guide Dr. L. R. Tiwari, Principal (Rtd.), M.D.College, Parel, Mumbai for his inspiring and encouraging way to guide for understanding of knowledgeable work, and his invaluable comments during the entire work towards the completion of this book.

This book has been kept on track and been seen through to completion with the support and encouragement of number of people including my parents, family members, well wishers, my friends, colleagues and various institutions. It is a pleasant task to express my thanks to all those who contributed in many ways to the success of this study and made it an unforgettable experience for me.

I am also thankful to Dr.(Mrs.) A. S. Kulkarni, Principal, N.B.Mehta Science College, Bordi, for her continuous support, motivation and encouragements as well as corrections in book.

I specially thank to Mr.Vijay Kini, Reliance Infrastructure Ltd., Dahanu for his kind help and moral support. I also want to thanks Ms. Paramate Honey Visweswara Rao Sy. B.Sc. Microbiology, Student of N.B. Mehta Science College, Bordi for her sincere support.

Finally, I wish to express my gratitude to my parents, brothers, Mr. Anil S. Kadam and Mr. Kishor S. Kadam, my wife Mrs. Deepmala S. Kadam and sons master Parth and Yatharth for their constant unconditional support and patience throughout my research work, without which this work would not have been possible.

Author

Dr. Surendra S. Kadam

Acknowledgement

I, Dr. Debasish Mahapatro first author of this book want to dedicate this piece of work to my father Late Suresh Chandra Mahapatra and mother Banaja Mahapatra. I am also thankful to my Brother Abinash, Asutosh and uncle Mr. Jogeswar Choudhury for their support.

I express my love and compassion to my wife Smt. Geetanjali Mahapatra, daughter Ms. Upasana Mahapatra and niece Aradhana Mahapatra. I also want to thanks Mr. Jubaraj Satapathy and Smt. Suramaa Satapathy for their sincere support and encouragement to complete the work.

I express my deep sense of gratitude to my Ph.D. supervisors Prof. R.C. Panigrahy and Dr. Sudarsan Panda for their intense help, advice and scholastic guidance to complete my doctoral work. I also wish to express my gratefulness to Prof. Lakshman Nayak, Head of the Department of Marine Sciences, for providing the necessary facilities to carry out this work.

The scientific staff of Wetland Research and Training Centre of Chilika Development Authority, Badkul, Khordha for their assistance to execute the field sampling program is highly appreciated.

I also want to convey my sincere gratitude to the Director, of Zoological Survey of India and Project Director, World Bank Sponsored ICZM Project implemented in Odisha under Ministry of Forest and environment, Government of Odisha for providing me a chance to work in Chilika.

I deeply acknowledge the knowledge based support provided by Dr.Rajani Kanta Mishra of NCAOR, Goa and the scientists of Zoological Survey of India such as Dr. Basudev Tripathy, Mr. S.S. Mishra, Mr. Aruendra Mahapatra, Dr. Sameer Kumar Pati for their sincere cooperation for the accomplishment of this book.

Finally, I want to thanks all my friends and relatives who silently supported to complete this book in time.

I pray to almighty for their wellbeing.

Place: Brahmapur, Ganjam, Odisha

Date: 22nd January 2018 (Saraswati Puja) ***Dr. Debasish Mahapatro***

List of Tables

List of Figures

Abbreviation

ANI	:	Andaman and Nicobar Is.
CWPRS	:	Central Water and Power Research Station
cm	:	Centimetre
cm^2	:	Square Centimetre
CMFRI	:	Central Marine Fisheries Research Institute
CR	:	Critically Endangered
E	:	East
EIA	:	Environmental impact assessment
EOL	:	Encyclopaedia of life
et al.,	:	Others
GBIF	:	Global Biodiversity Information Facility
GoK	:	Gulf of Kachchh
GoM	:	Gulf of Mannar
ICAR	:	Indian Council of Agricultural Research
Is.	:	Island
IUCN	:	International Union of Conservation of Nature
km	:	Kilometre
kmph	:	Kilometre per hour
LI	:	Lakshadweep Island
$mg\ l^{-1}$	:	microgram per litre
MSIP	:	Marine Species Identification Portal
N	:	North

NIO:	:	National Institute of Oceanography
no.	:	Number
NT	:	Near Threatened
°C	:	degree centigrade
ppt	:	parts per thousand
sp.,	:	Species
VU	:	Vulnerable
WORMS	:	World Register of Marine Species

1

Introduction

Preamble

1.1 Definition of coastal lagoon

Coastal lagoons are special type of ecosystems formed along the coast lines. To explain in a better way, many definitions have been proposed by researchers for coastal lagoons. The first scientific definition appears to be that given by Emery & Stevenson (1957) which states that "*coastal lagoons are the bodies of water, separated in most cases from the ocean by offshore bars or islands of marine origin and are usually parallel to the coast line*". Pritchard (1967) had opined that coastal lagoons can be treated as a form of estuary. Hence the definition of a lagoon can be same as that of an estuary i.e. "*semi enclosed coastal bodies of water which has free connection with the open sea and within which the sea water is measurably diluted with the fresh water derived from land drainage*". This definition however does not include lagoons which are periodically or permanently cut off from the sea. Based on the research findings of numerous investigators, Phelger (1969) defined coastal lagoons as "*shallow inland marine waters usually oriented parallel to the coast, separated from the oceans by one or more number of restricted inlets*". Later, Lankford (1977) had defined the coastal lagoon as "*a coastal zone depression below mean high water, having permanent or ephemeral communication(s) with the sea, but protected from the sea by some type of barrier*". Subsequently, Day (1980), however had shared the views with Pritchard (1967) and stated that coastal lagoons are also a type of estuary. He defined coastal lagoons as "a *partially enclosed coastal body of water, which has either permanently or periodically opened into the sea and within which there is measurable variability of salinity due to admixture of seawater and freshwater derived from inland drainage*". This definition, indeed, includes a variety of coastal water bodies viz. estuaries, coastal lagoons, saline lakes, small inlets of the seas etc., even though they possess different geomorphological attributes. The latest definition of a coastal lagoon, that appeared in the American Geological Institute (AGI) Glossary of Geology (Bates and Jackson 1987), is "*a shallow stretch of sea water, such as sound channel Bay or a salt water lake near or communicating with the sea and partly or completely separated from it by a low, narrow or elongate strip of land such as a reef, barrier island, sand bank or spilt*". The latest and most updated definition of a coastal lagoon as given by Kjerfve (1994) however is a *shallow coastal water body separated from the ocean by a barrier connected, at least, intermittently to the ocean by one or more restricted inlets*".

This definition has been widely accepted by the limnologists, ecologist, policy makers and scientists for different purposes, related to EIA studies, policy implementation and management practice.

1.2 Classifications of coastal lagoons

Coastal lagoons are very dynamic and complex ecosystems in the coastal zone. The hydrobiology of a lagoon depends upon its geomorphology and local environmental conditions. Therefore, classification of coastal lagoons on a broad scale is difficult. Coastal lagoons, in general however have been classified into different groups by different scientists. Nicholas and Allen (1981) had classified the coastal lagoon into four categories namely (i) estuarine lagoons, (ii) open lagoons, (iii) partially closed lagoons and (iv) closed lagoon. Kjerfve (1994), on the other hand, he grouped the coastal lagoon into three types such as (i) choked lagoon, (ii) restricted lagoon and (iii) leaky lagoon. The characteristics of these three categories are as follows.

i. **Choked Lagoons:** A Choked lagoon (Figure 1A) usually has a very narrow connection with the nearby sea. Thus it normally possesses tidal oscillations reduced to 5% or less in comparison to the adjacent coastal tide. This type of lagoons are also characterized by slow flushing rate of water and sediment, high wind induced wave action, dominant wind forcing and irregular stratification of different environmental parameters under the influence of solar radiation and surface runoff from catchment areas. In arid or semi-arid region, the choked coastal lagoons remain permanently or temporarily hypersaline (Copeland 1967; More and Slinn 1984). These types of lagoons are found in association with river deltas or with the nearby shore line. The examples of such chocked lagoons are the Lagoa dos Patos and Lagoa de Araruama of Brazil, in Lake St. Lucia of South Africa, the Coorong of Australia, and Songkla Lake of Thailand.

ii. **Restricted Lagoons:** Restricted lagoons (Figure 1B) are large water bodies commonly found parallel to the shoreline. They possess two or more inlets connected with the nearby sea or ocean. This type of lagoons generally experiences good tidal circulation amplified with wind generated wave action. They represent well mixed water mass and low vertical stratification. Wide range of salinity variations ranging from near limnetic to oceanic condition are noticed in this type of lagoons. Flushing rate in this type of lagoon is faster as compared to the chocked lagoons. The examples of such restricted coastal lagoons are: the Laguna de Terminos of Mexico, and Lake Pontchartrain in USA and Chilika lagoon of Odisha, India (the present study area).

iii. **Leaky Lagoons:** Leaky lagoons (Figure 1. C) in general are elongated water bodies found parallel to the shore line. They have many inlet connections to the adjacent sea. The characteristic features of the Leaky lagoons are wider channels or inlets, strong tidal currents, strong littoral drift action, strong wind action and the salinity is somewhat similar to the nearby sea experiencing quick flushing rate. Examples of such lagoons are the Mississippi Sound in USA.

1.3 Factors affecting the ecology of coastal lagoons:

Being a coastal transitional water body, a number of factors of physical, chemical, geological and biological origin affects coastal lagoon ecology. Some of the significant factors associated with the coastal lagoon are the geology of catchment area, coastal geomorphology, plate tectonics, source of inputs of sediment, water and nutrients supplied by the rivers and adjacent sea (Phleger 1981; Allen *et al.,* 1981; Zimmerman 1981, Nichols and Boon 1994; Bird, 1993; Smith, 1993; Mahapatro *et al.,* 2013). Physical forcing like the wind velocity, tidal incursion and coastal process are also putting significant pressure thereby structuring the lagoon ecology *eg.* Chilika lagoon. Again, salinity of such coastal ecosystems is found responsible for the distribution, abundance and species richness of their flora and fauna. The salinity regime of coastal lagoons is maintained by the precipitation - evaporation ratio of the water body and tidal water incursion into the lagoon. Like many other coastal ecosystems, Chilika lagoon is shallow, where the sunlight touches the bottom. The depth is varied greatly and governed by inflow of sediment from catchment area, tidal excursion of sea water and fresh water out flux under monsoon influence (Mahapatro *et al.,* 2013).

1.4 Distribution of coastal lagoon in India

Coastal lagoons are unevenly and widely distributed from tropic to sub-polar regions occupying about 13 % of the total coast line on earth (Troussellier 2007; Mahapatro *et al.*, 2013), but are absent in Polar regions. Many other environments which are considered similar to lagoonal ecosystem are the mangrove Islands and barrier reefs (Sourina and Ricard 1976; Mahapatro *et al.*, 2013).

The coast line of India is endowed with a large number of sensitive marine ecosystems in which coastal lagoons are dominant. The other sensitive ecosystems of Indian coasts are estuaries, coral reefs and mangroves. Many of these coastal habitats have been designated as the hot spots of biodiversity and productivity.

There are 17 coastal lagoons (Figure – 2) found along the coast lines of India *(source: http: // www. casmbenvis.nic.in/%5Cdatabase_lagoons.htm)*. Eight coastal lagoons such as (1) Chilika lagoon, (2) Pulicat lagoon, (3) Pennar lagoon, (4) Bendi lagoon, (5) Nizampatnam lagoon, (6) Muttukadu lagoon, (7) Muthupet lagoon, (8) Gulf of Mannar lagoons are situated on the east coast, while the remaining nine lagoons such as (1) Vembanad lagoon, (2) Ashtamudi lagoon, (3) Paravur lagoon,(4) Ettikulam lagoon, (5) Veli lagoon, (6) Murukumpuzha lagoon (7) Talapady lagoon, (8) Lagoons of the Mumbai coast, (9) Lagoons of Lakshadweep are located along the west coast. Out of these 17 lagoons, the Chilika lake is the largest. It is also the largest brackish water lagoon in the entire Asia and second largest in the World (First being Lake Baikal).

1.5 Significant Coastal Lagoons of India

A lagoon is a shallow body of water separated from a larger body of water by barrier islands or reefs. Lagoons are commonly divided into coastal lagoons and atoll lagoons. Coastal lagoons formed along the gently sloping coasts where barrier islands or reef can develop off shore and the sea level is rising relative to the land along the shore. Because of their relatively low flushing rates, coastal lagoons are favorable habitats for primary producers like phytoplankton and aquatic plants. Nutrients are transported to lagoons from surface water and ground water flows and through exchange with ocean. In India there are four coastal lagoons having larger significance level . These are as follows:

(a) Pulicat lake

This coastal lake is located in north-east of Chennai between 13°24' and 13°47' N and 80°2' and 80° 16' E. It is the second largest brackish water lagoons in India. The water spread is about 350 km^2, mostly saline, shallow in depth and opened into the sea by a narrow channel located at the southern end of the lake. The lake proper is extending north and north-west direction. Width is maximum about 200 m and it is very shallow in nature (about 1 m depth). The lake mouth gets silted up and closed for two phases first is from June-July and second is in November-December. Salinity profile showed a large scale differences which ranges between 0.75 to 57.0 ppt i.e. brackish to hyper-saline condition. There is a manmade canal called Buckingham canal that moves the entire length of the sand spit which is also called as the "Sriharikota Spit" used for artificial satellite lunching operation, separates the lake-cum-lagoon from the sea. This canal is connected to northern mud flats. These navigational channel have supported for the colonization of an array of flora and fauna of lagoon type. This lake is much richer with fish diversity mainly constituted by Perciformes (56.82%), Clupeiformes (12.5%), Mugiliformes (7.95%), Atheriniformes (5.68%) and Tetraodontiformes (5.68%). The fish species composition reflects that most of them are of marine forms and very few belong to the fresh water habitat. Among Perciformes (56.82%) gobbids are observed as the single large dominating fish group gobioids (11.36%). The species *Nematalosa nasus* (21.54%) was recognized as the single dominant species which was frequently caught in every fishing attempt followed by *Hyporhamphus limbatus* then followed by Gerres *lucidus*, *Thryssa purava*, *Mugil cephalus*, *Sillago sihama*, *Oxyurichthys microlepis* and *Triacanthus biaculeatus*. Post monsoon period showed maximum encounter of species and this is the period when salinity is moderate.

The benthic habitat of Pulicat can be broadly catagorised into three zones namely (i) sandy with a little admixture of mud in south, (ii) entirely muddy zone in northern side and (iii) admixture of sand and mud that helps preponderance of an array of benthic communities. The "Sriharikota Island" is served as an important destination of migratory birds. However, this lake ecosystem is considered as one of the major congregation site of the some rare species of fauna and flora in India.

(b) Gulf of Mannar (GoM)

It is a large shallow bay forming part of the Laccadive Sea connected between the southeastern tip of India and the west coast of Sri Lanka, in the Coromandel Coast region. There is an Island called as Mannar Island that separates the Gulf of Mannar from Palk Bay, which lies to the north between India and Sri Lanka. During the year 1986, 21 islets groups lying off the Tamil Nadu coast between Thoothukudi and Dhanushkodi was declared the Gulf of

Mannar Marine National Park and declared a Biosphere Reserve in 1989. The Gulf of Mannar provides the habitat support more than 3,600 species of flora and fauna and considered as one of the biodiversity hotspot in Asia and in the world. A total of 117 hard coral species already recorded till date. Many of the endangered and vulnerable species of the world find their ultimate destination in the gulf of Mannar namely sea horses, sea cows, sea turtles, sharks and whales dolphins etc. The salinity profile of GoM is somewhat controlled by both the impacts of Bay of Bengal and by Arabain sea. Generally salinities of the Bay of Bengal are lower (35 psu) than the Arabian Sea which is high up to 36.5 psu. The seasonal difference of sea surface salinity around the Island is about 2 psu in the islets of GoM. Once upon a time, the Dugong was abundantly found in the GoM but it has not been seen as frequent as in the past. But excess of maritime activities have damaged it's natural habitat. Another interesting fact is that all species of marine turtles are seen in the GoM namely Oilve Ridley (*Lepidochelys olivacea*) which is very common followed by Green turtle (*Chelonia mydas*), Loggerhead (*Caretta caretta*) and Leatherback (*Dermochelys coriacea*) are recorded more or less. It has been observed that the GoM provides an ideal route for the migration of Olive Ridley sea turtles from Indian ocean and south Asia.

Mollucan fauna are generally seen on the sandy or muddy areas in the GoM. These are found in the pearl Banks and Palk Bay. The most common and economically important species which also called as the sacred chank is *Turbinella pyrum*. Edible oyster (*Crassosstre madrasensis*), *Madiolus auriculatus*, *Gafrarium tumidum* and *Anadara antiquata*, the window-pane oyster (*Placuna placenta*) are predominant among other available species.

GoM is the best habitat for the sea grass diversity belonging to the genus Thalassis, *Halophila*, *Halodule*, *Enhalus* and *Cymodocea* are common. But the most dominant genus are those of *Thalassia* and *Syringodium* those are predominate in the coral reef areas as well as in the muddy and fine sandy soils. Many of the true mangrove vegetations are marked such as *Lumnitzera racemosa*, *Sonneratia casseolaria*, *Avicennia marina*, *Pemphis acidula*, *Rhizophora mucronnata*. GoM contains one of the rich coral reef diversity and reef associated fish and shell fish diversity of the world which needs proper conservation steps to protect its biodiversity.

(c) The Chilika Lake

The Chilika lake is the largest brackish water coastal lagoon ecosystem in Asia situated on the east coast of India, adjoining to the Bay of Bengal. It is one of the finest repositories of floral and faunal diversity, designated as a Ramsar

Site in India in the year 1981. The first systematic research on coastal lagoon water quality and fauna in India was probably started from Chilika lake in early parts of 20th century by researchers like Annandale (1915); Kemp (1915); Annandale (1915-1924), Sewell and Annandale (1922). They have described the hydrography, faunal assemblage and fisheries of the lake quite elaborately. Thereafter also several studies have been undertaken leading to piling up of literature dealing with topography, origin, water quality, flora and fauna including plankton & fisheries, sediment characteristics and environmental issues of recent origin as evident from the three successive review made by (Mahapatro 2017). The results of the investigations as elucidated in different publications relating to the Chilika lake suggests that its climate, sediment distribution, geomorphology, hydrography, water chemistry and biodiversity has experienced several changes over the years under the influence of several natural and anthropologic factors of both, the marine and terrestrial origin (Mahapatro 2017)..

(d) Vembanad lake

It is the longest lake in India and largest in Kerla district. It is included in the list of Ramsar Convention. It is the major destination for the migratory birds more than 20,000 waterfowls. It was originally a part of the shallow coastal area of the Arabian Sea. Because of the geological uplift it had formed a shallow bay like structure in which fresh water drained rivers rivulets channel are drown to it. There is a rich bed of mangroves mostly the species of *Rhizophora apiculata*, *Derris heterophylla*, *Sonneratia alba*, *Acanthus ilicifolius*, *Acrostichum aureum* and *Cerbera manghas*. Because in the natural anthropogenic activities a major chunk of mangrove forest is destroyed over the years of time. But, some of the patches are still sustained in few of the isolated pockets comprised of the genus *Rhizophora* and *Sonneratia*. There are large number of mangrove associated plant species found namely *Acrostichum aureum*, *Ardisia littoralis*, *Avicennia officinalis*, *Barringtonia racemosa*, *Bruguiera gymnorrhiza*, *Cerbera odollam*, *Clerodendrum inerme*, *Derris trifoliata*, *Dolichandrone spathacea*, *Excoecaria agallocha*, *E. indica*, *Flagellaria indica*, *Heritiera littoralis*, *Hibiscus tiliaceus*, *Kandelia candel*, *Morinda citrifolia*, *Pandanus fascicularis*, *Pongamia pinnata*, *Premna serratifolia*, *Rhizophora apiculata*, *Stenochlaena palustris*. The Vembanad Lake is rich in faunal and floral biodiversity and can be comparable with any of the other coastal ecosystems of India.

1.7 Problems associated with coastal lagoons

Being the transitional water bodies between the land and sea, coastal lagoons are ephemeral and fragile ecosystems susceptible to rapid environmental degradations imposed both by natural calamities and manmade activities. These ecosystems often face many problems causing pollution and biodiversity loss. In recent decades environmental stress belonging to natural and anthropogenic origin on coastal lagoons are on rise in day-by-day and several coastal lagoons have been recognized as hot spots. A few of the current problems prevailing in the coastal lagoons are reduction in salinity or salinity extreme condition, deposition of silt and reduction in depth, over exploitation and pollution (Anthony *et al.,* 2009 and Mahapatro *et al.,* 2013). The Chilika lagoon which is the best example of a restricted coastal lagoon too has been exposed to environmental degradations as explained in the reviews of Panigrahy (2000) Panda *et al.,* (2008) and Sahu *et al.,* (2014). The most noteworthy degradations were however the salinity reduction, siltation and eutrophication that had promoted the rapid growth and proliferation of noxious weeds belonging to submerging, surface floating and emergent types. In recent years, the emerged weeds like *Phragmites karka, Scripus sp,* surface floating weed like *Eichornia, Hydrilla* and submerged weed *Potamogetone pectinatus* have emerged as the important threat to the Chilika lake environment. The weed infestation coupled with the increased chemical pollutant load has significantly affected the health of this ecosystem causing biodiversity loss. In view of such problems, it is granted necessary to monitor the environmental status of this dynamic ecosystem with an efficient indicator. Macrobenthic organisms have been measured as most suitable organisms to indicate the environmental stress due to their sessile mode of life and longevity. A number of researchers' have advocated for using macrobenthic fauna as the environmental indicators of coastal embayment especially to study the environmental degradation arising due to pollution or other types of natural and anthropogenic impacts as documented by Sahu *et al.,* (2014) and Mahapatro (2017).

1.8 Macrobenthos

The community of organisms which live on, in, or near the floor of an aquatic environment is commonly known as benthos. This community is always associated with the sedimentary environment of lakes, estuaries and oceans. The term benthos was first coined by Haeckel in 1891 that refers to the Greek word '*benthos*' meaning the '*depth of the Sea*'. Organisms of this community lives in or near to the sedimentary environments of lakes and oceans extending from intertidal shore lines to deep abyssal depths. They have been classified in many ways depending upon their size range, distribution in sea floor, mode of nutrition etc. Mare (1942) classified the benthic organisms into three groups with respect to their body size. The organisms, which cannot penetrate through the mesh size of 500 micron are known as macrobenthos, whereas those organisms which can penetrate the mesh size of 500 micron but retained over 63 micron mesh are called as meiobenthos and the organisms that could able to penetrate the 63 micron mesh are called as microbenthos. This classification is widely used in size range classification of benthos (Holme and McIntyre, 1971; Alongi, 1990). According to their distribution in sedimentary environment, the benthic animals are further divided in to two major groups namely 'epifauna' and ' infauna'. The term "epifauna" is applied to those benthic organisms which live on the surface of the sea bed such as on rocks, stones, shells and other sediment particles. But the term 'infauna' is used for those which can live either as buried or burrowing animals inside the sediment. The former group is dominated by whole block of phyto community like seaweeds, sea grass and several groups of animals like coelenterates including corals, sponges, sea anemones, bivalves, barnacles, sea star, brittle star, and many crustaceans commonly known as zoobenthos.

The epifaunal community constitutes about 4/5th of total bottom living invertebrates on the sea floor (Holme and McIntyre 1971; Alongi 1990; Mahapatro 2017). Macrobenthos comprises relatively larger, more visible, animals living on the floor of the aquatic milieu. They include a wide range of invertebrates, corals, sponges, polychaetes worms, bivalves, echinoderms, sea anemone, sea squirts turbellarians and several larger crustaceans like crabs, lobsters, and cumaceans. The ecological and economic importance of benthos especially the macrobenthos is well established. Many of these organisms serve as scavengers by eating organic detritus that sunk to the bottom from surface and / or entered through surface runoff. Few species, especially the soft bottom polychaetes and mollusca are treated as good indicators of organic pollution and heavy metal pollution along the coastal waters of the world. Molluscs comprised of both bivalves and gastropods play a significant role by converting complex food materials to their simpler forms. Therefore, supporting the bio-geo- chemical cycling. Most of the mollusca are suspension feeders hence, they accumulate planktonic materials, microbes with other substances in to their gills and later on these are used for its body metabolism. Since, molluscs are predated by a number of predators such as, fishes, shell fishes, birds and human being. Thus, they make an important link between benthic-pelagic trophic relation and energy transfer. It also helps to maintain the bottom soil fertility of any aquatic ecosystems. Since, some of the molluscs feed on the dead organism matters hence they keep the bottom habitat clean. Many of the molluscan species are used in the eco-toxicological studies, in EIA studies etc. Some of the molluscs contains bioactive compounds those have major role in the medicinal and pharmacological industry. Dead shells of the mollusc are used for decoration purposes, as medicines for the preparation of lime etc. Considering their significant role of mollusc, the present study is designed to describe the molluscan diversity of Chilika lake during the year 2007 to 2010.

1.9 Mollusca:

Mollusca is one of the largest phylum in the animal kingdom. They are the mostly sedentary and posses' very weak power of locomotion. The words mollusc and mollusk both are derived from the French *mollusque*, which originated from the Latin *molluscus*, from *mollis* that means soft. However, the word *Molluscus* itself was an adaptation of Aristotle's *ta malaká*, that means "the soft things", which had been named to cuttlefish. The scientific study of molluscs is called malacology **(**Little *et al.*, 1964**).** Substantial evidence exists in the form of fossilized samples to explain that the origin of gastropods, cephalopods and bivalves are in the Cambrian period i.e. about 541 to 485.4 million years ago. This phylum is undoubtedly diversified with an array of body shape, size, behavior, habit and habitat. For example giant squid of the class Cephalopod is regarded as the largest invertebrate in the world that has well developed neurological advancements. Similarly, the class gastropod contains largest number of identified species among the other molluscan groups.

There are many features to explain the molluscan characteristics. However, three conspicuous features are considered the significant. Those are (i) containing a mantel cavity helpful for breathing and excretion, (ii) presence of a radula (except for bivalves), and (iii) a well defined nervous system. Phylum mollusca include 23% of the total identified benthic invertebrate species globally, habituated to live in the marine, brackish water, terrestrial and fresh water forms. But, a major bulk is remained as marine form. Starting from intertidal area to deep abyssal depths molluscan groups are available in large in terms of abundance and diversity. About 80 % of the total molluscs are belonging to the class gastropods (Ponder& Lindberg 2004) followed by bivalves. Globally there is a haphazard understanding about the total species of mollusk species existing on earth today. According to a rough estimation, the total number of living species near about 50,000 to 1,20,000. Chapman (2009) presumed that the total number of living species is 85,000 while Haszprunar (2001) described the total living and identified species is about 93,000 out of which 23% of all named marine organisms. In general if we consider the findings of Chapman (2009) and Ponder& Lindberg (2004) then the number may be of 1,20,000 of identified living molluscan fauna recorded till date and if about 70,000 fossils forms are added subsequently the total number is near about 2,00,000 (Brusca & Brusca (2003).

Now days, there are five classes considered by the taxonomist fall under this phylum namely (i) Polyplacophora, (ii) Gastropoda, (iii) Bivalvia, (iv) Scaphopoda and (v) Cephalopoda. In Indian context, mollusks have been observed from rocky coasts, sandy beaches, muddy shores, intertidal mudflat, continental shelves, sea grass beds, mangroves & coral reef ecosystems and at abyssal depths in the sea/ocean. Globally, there are 586 families are recorded out of which 279 families are exclusively found in Indian sub-continental region comprised of 3600 species from which about 65% (2300) are exclusively the marine forms.

1.10 Molluscan study in India:

The first scientific study on mollusca was started towards the beginning of 20th centaury in Chilika lake by the researchers like Preston, Annandale, Kemp, Eliot, Sewell, Hornell, Godwin-Austen, Prashad etc. Then the molluscan taxonomical research had got momentum by the workers like Adam(1939), Ray (1948a,b, 1952, 1954), Hornell and Tomlin (1951), Subrahmanyam *et al.,* (1952a&b), Patil (1954), Satyamurti (1956), Patil and Gopalkrishnan (1960), Tikader (1964), Cheriyan (1968), Narayanan (1968, 1969, 1972), Prabhakar Rao (1968), Subba Rao (1968, 1970, 1980, 2003), Joshi (1969), Radhakrishna and Ganapati (1969), Nagabhushanam and Rao (1972), Satyanarayana Rao and Sundaram (1972), Virabhadra Rao and Krishna Kumary (1973, 1974), Rajagopal and Subba Rao (1974), Starmuehlner (1974), Virabhadra Rao *et al.,* (1974), Subba Rao and Dey (1975, 1984, 1986, 2000),

Subba Rao and Mookherjee (1975), Das *et al.,* (1977, 1981), Kohn (1978), Rajagopal and Mookherjee (1978, 1982), Namboodiri and Sivadas (1979), Gopinadha Pillai and Appukuttan (1980), Subba Rao and Surya Rao (1980, 1991, 1993), Subba Rao *et al.,* (1983, 1987, 1991, 1992, 1993, 1995), Kasinathan and Shanmugam (1985), Mookherjee (1985), Tikader and Das (1985), Tikader *et al.,* (1986), Jothinayagam (1987), Mookherjee and Barua (1989), Apte (1992, 1993, 1997, 1998, 2009), Babu Philip and Appukuttan (1995), Surya Rao and Maitra (1998), Mahapatra (2001, 2008, 2010), Surya Rao *et al.,* (2004), Venkataraman *et al.,* (2004), Subba Rao and Sastry (2005), Dey *et al.,* (2005), Dey (2006, 2008), Arularasan and Kasinathan (2007), Ramakrishna *et al.,* (2007, 2010), Rao and Sastry (2007), Roy *et al.,* (2008), Apte *et al.,* (2010),Raghunathan *et al.,* (2010), Rao (2010), Venkitesan and Mukherjee (2011), Mahapatro *et al.,* (2015), Mahapatro *et al.,* (2016), Mahapatro (2017). In addition to this several information are published in the form of conference proceedings and unpublished reports, as dissertations and Thesis which is found difficult etc to incorporate all.

1.10.a Molluscs of East coast of India

A total of 2199 species of molluscs under 588 genera and 185 families have been recorded from the east coast of India. Polyplacophora having 117 species comprised under 12 genera and 6 families; Gastropoda of 1487 species of under 344 genera and 104 families; Cephalopoda 54 species of under 23 genera and 11 families, Bivalve having 632 species of under 207 genera and 63 families; Scaphopoda 9 species under 2 genera and one family as described by Venkataraman *et al.,* (2012). However, this is doubtful and need proper verification before considering the total number of species in to the account (Tripathy and Mukhopadhyay 2014). Available literature suggested 3 species of cephalopods, bivalves of 36 species, gastropods of 25 species, 1 each of scaphopoda and polyplacophora are endemic to east coast of India. Nevertheless, *Tudicla spirilus* is endemic to the Bay of Bengal and east coast of India. This species of gastropod is also included as scheduled mollusca in the Wildlife Protection Act (1972) of the Government of India.

There are many expanses in the east coast of India where huge congregation and diversity of mollusk can be seen. These are Sundarban mangroves, Bakkhali, Digha, Talsari Beach, Subarnarekha estuary, Dhamara estuary, Chandipur, Mahanadi estuary, Outer channel and main lake area of Chilika, Ruskikulya estuary, Chatrapur coast, Gopalpur coast and Bahuda estuary.

In south Bimunipatnam and Visakhapatnam up to Konada, the Kakinada Bay, Godavari and Krishna river estuaries, Pennaru estuary, the Pulicate lake, Gulf of Mannar and Palk Bay, Nagapattinum, Tuticorin and Kannyakumari coast are the conspicuous locations of molluscan communities.

There are many species of mollusc available in east coast of India as threatened species because of the exploitation as food, as medicine, as decorative purposes as ritual believe etc. As an example Windowpane oyster *Placuna placenta* not only is a best source of lime but also it muscle part is edible, and the right valve is used for the decorations in the house, in the glazing windows (Menon and Pillai, 1996). A sacred chank *Turbinella pyrum* is now becomes vulnerable because of the massive exploitation due to its high market value. Once upon a time it was available in huge in the south India but now occurrence of this species becomes rare. Other molluscan species such as *Harpulina arusiaca, Fasciolaria trapezium, Cypraea limacine, Tudicla spirillus, Conus milnedewardsi, Cassis cornuta, Nautilus pompilus* are under threat since of their commercial value in the open market. Thus large scale exploitation helped to decline their population.

1.10.b Molluscs of west coast of India:

Like east coast, the west coast of India is similarly rich in the mangrove biodiversity, coral reefs, mud flats, rokey shore, Islands etc. But there is significant drop down in the entire species number of mollusca in the west coast. Perusal of published literature revealed that a total of 1046 molluscan species are recorded till date in contrast 2199 species as recorded in the east coast of India. Out of these 1046 species 707 species of gastropods, 248 species of bivalves, 2 species of scaphopoda, 80 species of cephalopods, and 9 species of polyplacophora occur along the west coast. Actual reason of lower abundance and biodiversity of mollusk in the west coast of India might be many but it is believed that lesser attempts regarding field surveys or it could be the consequence of long term anthropogenic activities since the west coast is richer with industrial set ups, oil rigs, petrochemical refineries, port activities.

Since the west coast of India has a good mangrove vegetations, coral reef ecosystem, the Gulf of Kutchch, Malvan, estuarine ecosystems of Mandvi, Juari, Narmada, Tapi, many of the backwaters in Kerala coastal sandy stretches of Karnataka namely Gangoli, Kundapur, Kerala Alapuzha, Kovalam are observed as the rich ecosystem of the molluscan faunal communities.

Along the west coast of India certain species are found as endemic such as 35 species of gastropods, bivalves having 8 species recorded as endemic. But no species were found endemic belonging to the group scaphopoda, cephalopoda or opisthobranchia. More study is required to carry out although very less attention has been paid so far. Similarly, many species were threatened in the west coast of India. Many species were reported again once they have recorded for the first time or after their discovery. Besides this, like east coast, west coast also exhibited massive exploitation of *Placuna placenta* for the extraction of lime and decorative purposes. Similarly, large scale exploitation of *Meretrix casta*, is now recorded by the local fishermen for its nutrition value. Other mollusks such as *Strombus plicatus* Sibbaldi, *Nautilus pompilius*, *Paphia malabarica*, *Villorita* sp., were exploited in the coastal districts of the Gujrat, Maharastra, Goa, Karnatak etc. the reason of the exploitation is largely for the nutritional purpose, followed by decorative desires. Most of these species mentioned above are included as schedules species of Wildlife (Protection) Act, 1972.

1.11 Prominent coastal ecosystems in India enriched with Mollusca

Andaman and Nicobar Is. (ANI) –

According to the published reports, about 1147 species of mollusca were recorded till date from 384 genera and 143 families.. Polyplacophora is represented by 12 species, Cephalopoda is comprised of 33 species, bivalves of 350 species, class Scaphopoda is represented by 7 species. There are about 177 species endemic from which 49 from gastropods and 18 bivalves from bivalve. ANI also contains a bunch of molluscan species those are also recorded from Indo-Pacific region. A total of 45 species exclusive of Indo-pacific origin are observed from the ANI. A species named *Charonia tritonis* has restricted distribution in Nicobar only, while in Andaman *Turbinella pyrum fuscus* is conspicuous. The conditions of ANI is found to be suitable for the substance of such endemic species. Many of the threatened species are seen in the ANI such as top shell *Trochus niloticus,* Turban shell *Turbo marmoratus,* giant clam *Tridacna maxima,* trumpet shell *Charonia tritonis etc.* The turban shells of ANI are on the edge of extinction due to over exploitation for human consumption (Raghunathan *et al.,* 2010).

Lakshadweep Island (LI) –

Nagabhushanam and Rao (1972) found that a total of 191 species of marine mollusca under 94 genera were predominated in the LI. Out of these, Polyplacophora contains 3 species belong to 3 genera; Gastropoda of 130 species belong to 67 genera;, Cephalopoda of 7 species belong to 5 genera, and Bivalvia of 51 species belong to 19 genera in the Lakshadweep Is. Later, Surya Rao and Subba Rao (1991) provided an extended checklist of mollusks from Lakshadweep islands which is belonging to 424 species from 201 genera and 105 families. Polyplacophora represented by 4 species of under 3 genera and 1 family, gastropods of 303 species under 138 genera and 70 families, cephalopods of 11 species of under 7 genera and 5 families, and 107 species of bivalves under 53 genera and 29 families were recorded.

Gulf of Mannar (GoM)

According to Satyamurti (1952) Krusadai Island and Gulf of Mannar, a total of 258 species of polyplacophora and gastropoda of which 8 species of polyplacophora are under 3 families were recorded. While gastropods of 250 species belonging to 63 families were recorded. Some of the species of bivalves dominating the GoM marine reserve are those of *Arca conplanta, Branchidontes variabilis, Chama reflexa, Gastrochaena gigantean, Isognomon nucleus, Lithophaga teres, Lithophaga gracilis Lithophaga nigra, Lithophaga cinnamomea, Modiolus metacalfei, Ostrea* sp, *Petricola lithophaga, Pteria chinensis, Septifer bilocularis, Venus reticulata, Venerupis macrophylla, Crassostrea cuttackensis, Saccostrea cucullata, Lopha folium, Anodontia edentula, Cardites antiquata, Cardita varieguata, Trachycardium flavum Cardium coronatum, Tridacna maxima, Solen lamarckii, Cultellus maximus, Tellina ansulata, Macoma ala, M. bruguiera, Placuna placenta, Volachlamys tranquebarica, Chlamys crassicostata, Spondylus imperialis, Decadopecten plica, Pinctada vulgaris, Pinna atropurpurea, Modiolus metcalfei, Perna viridis* etc.

Among gastropods the species such as *Haustellum haustellum, Littorina scabra, Nassarius papillosus, Nerita albicilla, Nerita chameleon, Nerita Maura, Nerita rumphi, Nerita squamulata, Phalium canaliculatum, Pyrene versicolor, Rissoina clathrata,* Strombus gibberulus, *Strombus canarium, Trochus radiatus, Trochus stellatus, Trochus tentorium, Turbo intercostalis, Turritella acutangula, Turritella attenuata, Vermicularia* sp. are predominate.

Marine Mollusca of Sundarbans Mangroves and Biosphere Reserves

According to Dey (2006) who worked on mangrove associated molluscs of Sundarbans, recorded 56 species of molluscs, under 42 genera and 30 families. From this, 31 species are gastropods belonging to 20 genera and 15 families and rest 25 species were reported from bivalve community.

Predominate species of gastropods are those of *Onchidium tenerum, O. tigrinum, O. typhae, O. verruculatum, Ellobium (Auricula) gangeticum, E. aurisjudae, E. (A.) translucens, Cassidula* sp., *C. nucleus, Melampus pulchellus, M. singaporensis, Pythia plicata, Cuthona Annandale, Pugilina cochlidium, Nassarius (Aciculina) subconstrictus, N. ennurensis, N. stolatus, N. orissaensis,* Cerilhidea obtusa, *Cerithidea cingulata, Thiara (Mainwaringia) paludomoides, Assiminea beddomeana, A. brevicula, A. jrancesiae, A. microsculpta, A. hungerfordiana, A. theohaldiana, A. woodmansonia, Gangetica miliacea, Stenothyra blanfordiana, S. deltae, S. gangetica, S. woodmasoniana, Litlorina melanostoma, L. carinifera, L. undulata, L. scabra.* Among bivalves *Modiolus strialulus, M undulatus, Enigmonia aenigmetica, Crassostrea gryphoides, C. palmepes, Saccostrea cuculata, Meretrix meretrix, Dosinia (Sinodia) trigona, Pelecyora trigona, Paphia undulate, Donax* incarnatus, *Theora opalina, Tellina (Pharaonella) iridescens, Macoma birmanica, M. qubemaculum, Strigilia (Aeretica) splendid, Glauconome cerea, Solen kempi; Tanysiphon rivalis, Siliqua* albida, *Pharella javanca, Corbula* gracilis, *Martesia striata, Bactivnophorus thoracites, Loligo indica, Octopus rugosus* are predominant.

Marine Mollusca of Gulf of Kachchh (GoK):

The basic information on the molluscan biodiversity was described by Subba Rao and Sastry (2005) who enlisted 178 species of molluscs under 123 genera and 72 families, including 1 species of polyplacophores and 124 species of gastropods under 85 genera and 52 families, 3 species of cephalopods under a single genus and family, and 50 species of bivalves under 36 genera and 18 families. Some of the dominant species are those of *Cellana radiate*, *Trochus radiates*, *Monodonta australis*, *Turbo intercoastalis*, *Turbo coronetus*, *Nerita albicilla*, *Cerethium caeruleum*, *Rhinoclavis sinensis*, *Cyprea lynx*, *Tibia insuladchorab*, *Thais bufo*, *Murex bruneus*, *Purpura panama*, *Cantharus spiralis*,, *Mitra sculata*, *Oliva oliva*, *Conus miliaris*, *Siphoneria siphoneria*, *Aplasia oculifera*, *Chiton peregrinus*, *Mytilus edulis*, Paphia *ala-papilionis*, *Sunetta scripta*, *Venus reticulate*, *Dosinia cretacea*.

Dhamara estuary of Odisha:

Dhamara estuary is more popular for the Dhamra port is located just 10 km away from the Gahirmatha estuary and Bhitarkanika marine sanctuary. A study made by Mahapatro (2006) during his M.Phil study he found 17 species of bivalves and 8 species of gastropods and most interestingly very less number of intertidal communities was observed. After 10 years another study was made whose data is unpublished yet. It showed very less representation of species diversity of mollusc at the later sampling program than the previous.

Bhitarkanika coast of Odisha:

A total of 238 species of marine molluscs have been reported from the Bhitarkanika coast of Odisha, including 1 species of polyplacophores. gastropods of 108 species under 65 genera and 53 families, cephalopods of 9 species under 6 genera and 4 families, 121 species of bivalves under 41 genera and 36 families, and scaphopoda of 1 species (Tripathy and Mukhopadhyay 2014).

Molluscan diversity of Mahanadi estuary:

A total of 149 species under 98 genera and 55 families. Of these 33 species (12 gastropods and 21 bivalves) were recorded by Surya Rao and Maitra (1998). Of the 11 species of gastropods except for *Onchidium tenerum* rest of all are marine and among bivalves 13 species are estuarine and 9 species are marine. The latter are represented by empty valves only.

Molluscan Diversity of Chilka lagoon:

When India was under the British colonial rules many of the British and Indian researchers have made significant contribution towards the molluscan research and these reports and collections are appeared as first baseline survey on Chilika. Few of those conspicuous studies are Preston during the year 1914 and 1915 had given a detailed picture of molluscan fauna of the Chilka Lake while in the subsequent year Annandale and Kemp (1916) descried about the occurrence of gastropoda and lamellibranchiata in the lake Chilika. In the next year Eliot (1916) reported on the nudibranchiata species from the lake premises. Immediately in the next year Godwin-Austen (1917) reported on the mollusk collected from the Barkuda island. During 1922 Sewell and Annandale, reported 20 species of mollusca from the Rambha bay. Annandale (1924) gave a further account on the gastropoda of the Chilka Lake. Prashad (1939) reported 2 species of mollusks. Molluscan faunal diversity in the Chilka lagoon is very rich and the same fact has been re-described by Subba Rao *et al.*, (1995) who observed and further documented 136 species of marine molluscs under 96 genera and 66 families from the lagoon. 59 species of gastropods under 41 genera and 30 families, 69 species of bivalves under 47 genera and 28 families, and a single family, genus and species of cephalopods were recorded during that study. Among the prevailing taxa few taxa were recorded all time dominant in the Chilika lake sediment. Those were Meretrix *meretrix, Meretrix casta, Neosolen aquaedulcioris, Macoma birmanica, Theora opalina, Clementias vatheleti, Modiolus undulates* and *Modiolus striatulus* were conspicuous. However, Chilika lake experiences large no. of endemic species and that measured about 39 % of the total taxa recorded during the year 1995. It has been observed that about 33 species of mollusca have their type localities in the Chilka lagoon. Recently, there is publication made by Mahapatro *et al.*, (2016) in which authors have summarized the findings of three years of in-depth study for a period of three years ranged from 2007 to 2010. The study documented 126 species of mollusc taxa belonging to 56 families, 18 orders and from 3 classes. Gastropod becomes dominanat having 64 species and 61 species were of bivalves followed by 1 species of Chiton. Outer channel area harbors maximum number of molluscan taxa as compared any other sectors covering the lake proper.

Among the predominate species *Meretrix meretrix, Meretrix casta, Neosolen aquaedulcioris, Macoma birmanica, Theora opalina, Clementias vatheleti, Modiolus undulates* and *Modiolus striatulus* among bivalves and *Cerethideopsilla cingulata, Bullia vittata, Nassarious stolatus, Indothais lacera, Natica tigrina, Turritella attenuate* were significant. Occurrence of a large number of marine taxa in the outer channel area is signifying that the lagoon ecosystem of Chilika gradually favoring marine taxa as compared to brackish water forms. This might be due to the opening of the new lagoon inlet during 23rd September 2000 and 1st August 2008.

Molluscan Diversity of Rushikukuya estuary–

Rushikulya estuary is well known as a mass nesting site for endangered olive ridley sea turtles. The molluscan community is of heterogeneous type but quite similar to that of the community observed in the Chilika lagoon located only 30 km northward from the Estuary. The study was carried out during years 2005-2006 and in years 2016-2017. A total of 67 species of mollusc taxa documented 35 species were from gastropod and 31 species of bivalves 1 species of *Chiton* during 2005-2006 while in 2016-2017 a total of 102 species were recorded and again gastropod becomes dominant having 61 species followed by the 40 species of bivalve and one species of *Chiton*. Some of the conspicuous species were those of *Meretrix meretrix, Meretrix casta, Macoma birmanica, Theora opalina, Modiolus undulates* and *Modiolus striatulus* among bivalves and *Cerethideopsilla cingulata, Bullia vittata, Nassarious stolatus, Indothais lacera, Natica tigrina, Turritella attenuate* were significant.

Scope and objective of the present study

This lake was exposed to severe environmental problems under the influence of natural calamities and manmade perturbations which has considerably threatened its biodiversity and fisheries (Panigrahy 2000; Panda *et al.*, 2008; Sahu *et al.*, 2014; Mahapatro 2017). The more common problems are the siltation, eutrophication, salinity reduction, changing of lagoon's inlets and increasing load of addition of pollutants etc having strong influence on the ecology. They together have facilitated the rapid infestation of noxious weeds that has worsened the situation in recent past. Thereafter, hydrological intervention was planned and based on technical advice a new mouth was cut opened at Shipakuda - south of old mouth. This has improved the water exchange between the lagoon and Sea there by increased the salinity condition, reduced the siltation, checked weed infestation and improved the overall biodiversity. The present study was therefore considered as ideal exercise to assess the influence of hydrological intervention on the molluscan community animals in the lagoon. The objectives of the present study include:

- To record the existing molluscan species of the Chilika lagoon
- To produce a pictorial guide on monograph of mollusca of Chilika

- To find out potential areas of molluscan abundance and diversity
- Conservation and Management

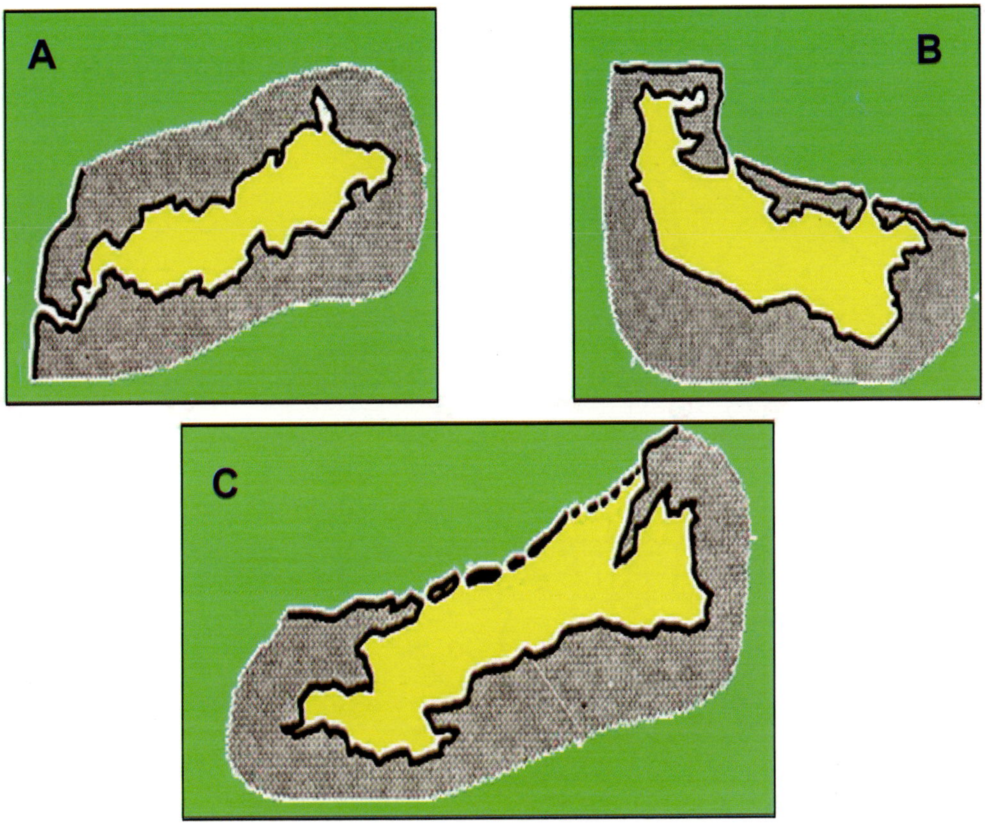

Figure 1: *Types of Coastal lagoons as proposed by Kjerfve, (1994), A. Choaked lagoon; B. Restricted lagoon C. Leaky lagoon.*

Figure 2: *Map of India showing the distribution of coastal lagoo ns on east coast and west. (Source: http://www.casmbenvis.nic.in/%5Cdatabase_lagoons.htm)*

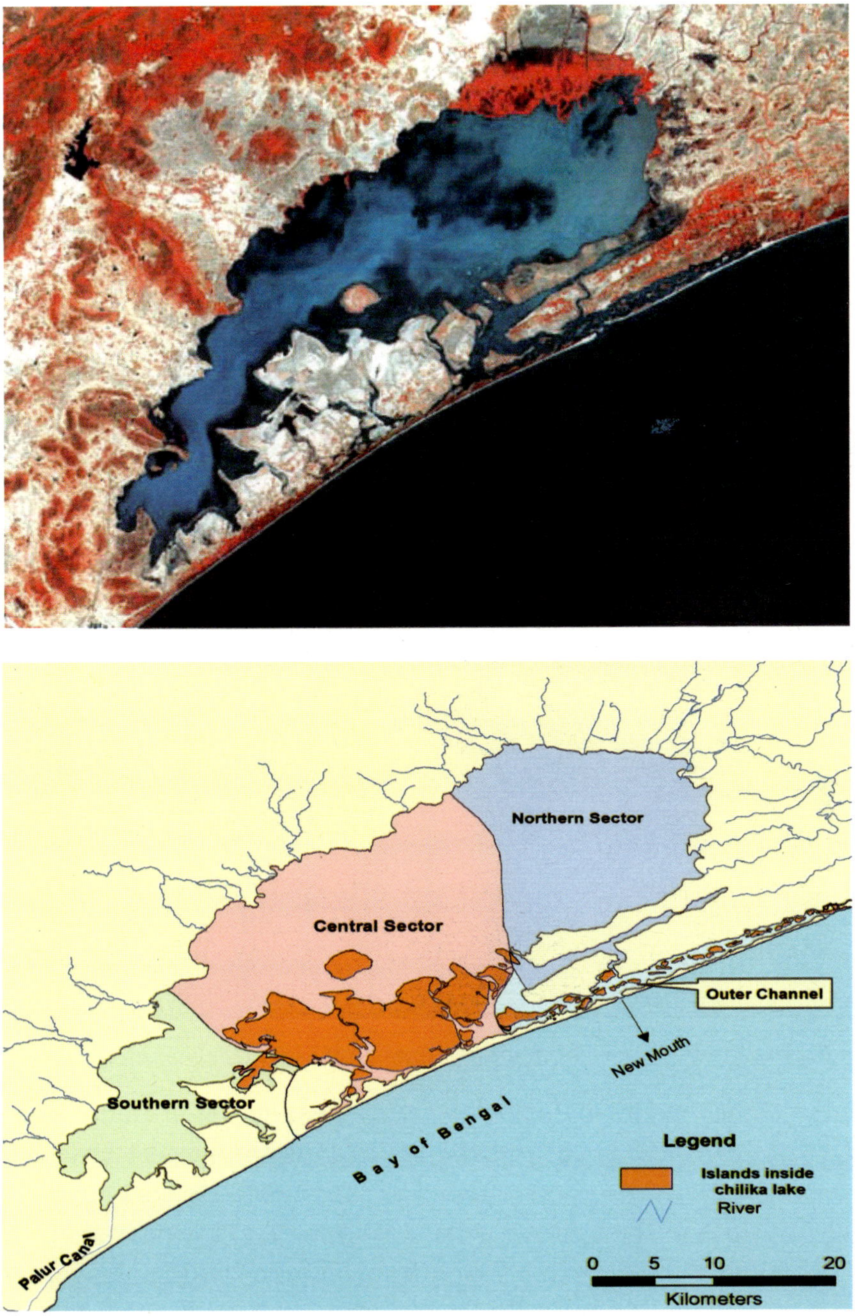

Figure 3: *Satellite image of of Chilika Lagoon (Upper) and sectoral division of lagoon (Lower)*

2

Study Area

Chilika lagoon

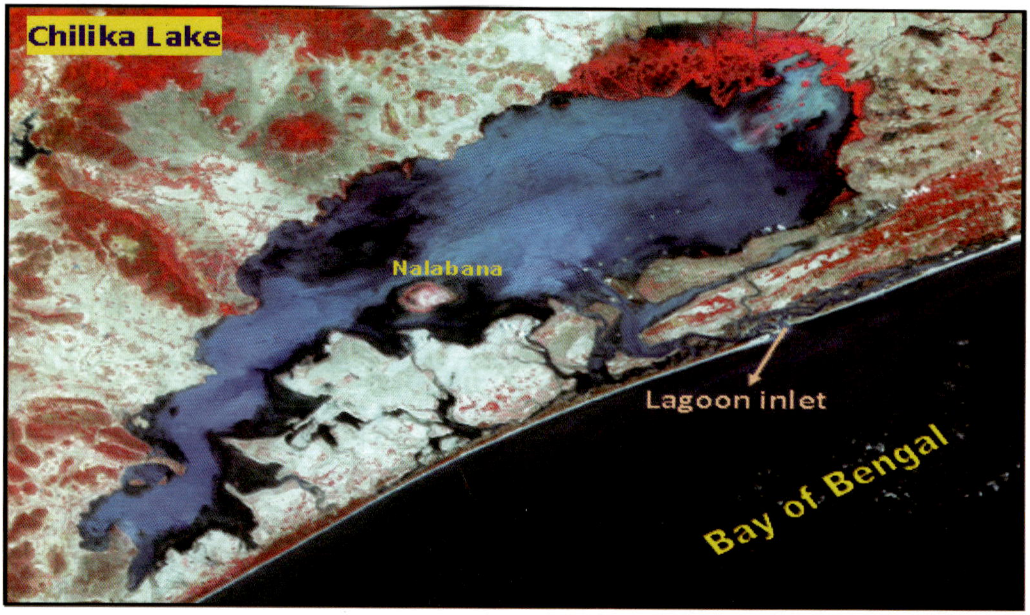

Figure 4: *Satellite Image of Chilika taken from IRS LISS III showing the lagoon inlet.*

Chilika lagoon of Odisha is located on the east coast of India between 19^0 28' and 19^0 54' N and 85^0 05' and 85^0 38' E. It is the largest brackish water wetland ecosystem in Asia is surrounded by the three districts namely Ganjam, Puri and Khurdha in the state of Odisha (Figure-4). This pear-shaped lake is about 64.5 km long and its width varies from 5 km to 20 km and average depth was one meter. This lagoon is shallow except some portions of the southern sector and outer channel area. Being shallow, the sunlight directly touches to the lagoon sediment. Hence, primary productivity is often high because of the growth of micro and microalgae, sea grass, sea weeds and other brackish water and fresh water weeds in huge. As a consequence, the lagoon provided a conducive condition for the fisheries production in large which is measured about 12000 MT per annum. About 0.2 million fishers living in and around it are directly depended up on it for to obtain their livelihood. OIt is world wide known as a biodiversity hotspot due to the coexistence of fresh water, brackish water and marine water sub-ecosystemsIt is alos the largest habitat for the migratory birds in India and as the largest habitat for Irrawaddy Dolphins may be next to the Sonkhala lagoon of Thailand. More than 120 number of Irrawaddy dolphins are seen in this lake as recorded in the recent years (Mahapatro 2017). The water spread area of the lake varies from 906 sq km in summer to 1165 in monsoon. It has been designated as Ramsar Site of international importance since 1981.

Origin

Chilika lagoon was formed around 6,000-8,000 years which has been proved by the many of the recent studies. So far the origin of Chilika lagoon is concerned; many theories are involved to explain. This has been accepted worldwide that this lagoon is formed due to the sea level rise in late Holocene period. From the available data and records regarding to the origin of Chilika might have originated during the pleostocene era, but the present form was gained after the Holocene period. The Geological evidences revealed that the coastline of Chilika was extended all along the western shores during the Pleistocene era, while the north-eastern region was under the sea. In due course of time, the coastline had moved eastward considerably and Chilika lake is fomed having one or more inlet connected to the sea. Synonymous event was found in case of Konark temple which was built on the seashore few hundred years before but it is now over 3 km far from the coastline. The shoreline of Chilika lagoon is much reduced due to the fluvial deposits of Mahanadi estuary in the northern flank of the lake area as shown in Figure 5 (Mahapatro 2017)

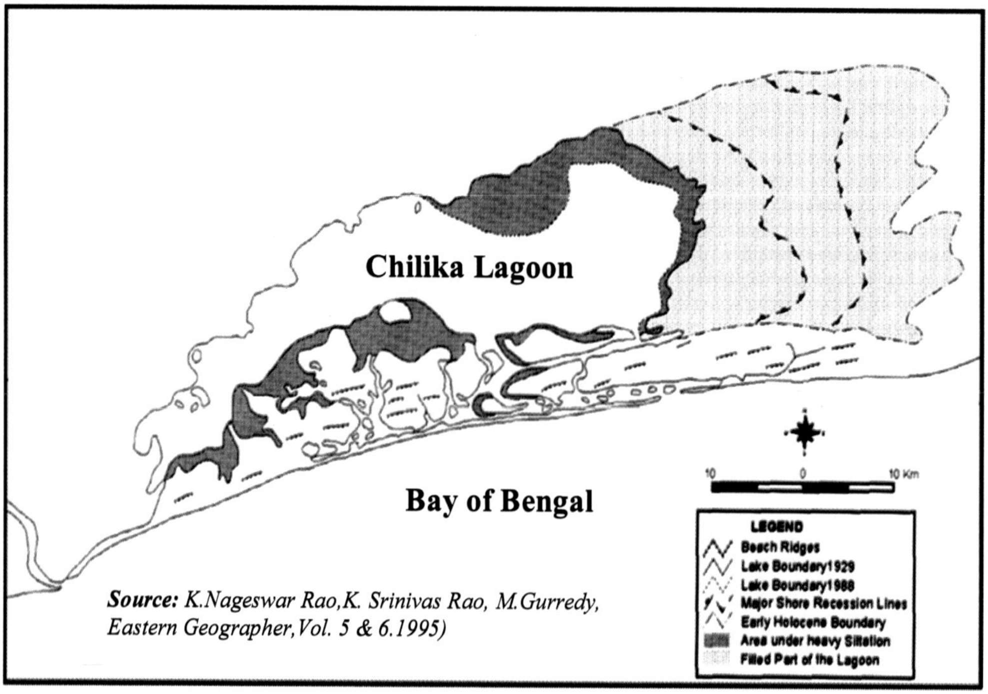

Figure 5: *Map Showing Changing Shoreline of Chilika Lagoon*

Ecological features of Chilka lagoon

Chilika lagoon is the most dynamic and highly productive ecosystem in the world. It is influenced by three hydrological units such as distributaries of river Mahanadi in the north, Palur channel in south and Bay of Bengal in the east. The distributaries Rivers of Mahandai namely Daya, Bhargavi, Luna and Makara carries silt borne fresh water to the lagoon ecosystem. A 16 kilometer long channel called "Palur channel" joins the river Rushikulya at south with the southernmost tip of the Chilika lagoon. On the eastern side, the lagoon is well connected with the Bay of Bengal by a couple of inlets through which tidal prism of the lagoon is maintained. The dynamism of the lagoon ecosystem is basically depending up on these three hydrological systems (i) Mahanadi delta, (ii) Westen catchment and (iii) Bay of Bengal in east where as the tidal rhythm is semidiurnal type. Ecologically, Chilika is an assemblage of freshwater, marine and brackish water ecosystems influenced by the intrusion and excursion of fresh water and sea water and their mixing. Lagoon is broadly classified in to four sectors such as Northern sector (fresh water), Central sector (Brackish water), Southern sector (Brackish-cum-marine) and outer channel area (marine zone except monsoon). Therefore, there is a well defined spatio-temporal variation in the physico-chemical parameters as well as biodiversity has been recorded.

Salinity becomes the main guiding parameter among others which strongly determines the distribution of flora and fauna of the lagoon. The salinity levels have shown marked annual variability, i.e. minimum in monsoon and highest at pre-monsoon seasons. Broadly, the salinity regime of the lagoon is controlled by precipitation & evaporation and water exchange between the sea and lagoon. The fresh water input commonly from northern and western catchment area during monsoon and intrusion of sea water through the inlets connected to the Bay of Bengal plays a major role for the maintenance of salinity. Salinity exerts considerable influence over the biodiversity of the lagoon. Though the water quality is brackish in nature, it is converted to a fresh water lake during monsoon due to massive ingress of fresh water into it. Other than this, wave dynamic due to wind and tide has also great influence in structuring the hydrography of the lagoon there by influencing its biodiversity. As many as 52 numbers of rivers and rivulets are opened into the lagoon poring fresh water. Major input from these rivers generally takes place from mid June to mid of October (monsoon season). There is a 32 km long outer channel which runs parallel with the Bay of Bengal separated by a sand bar endowed with two lagoonal inlets. The sea water enters in to the lagoon by the help of these two inlets or mouths that are opened to Bay of Bengal. The first was dredged on 23rd September 2000 near the village Sipakuda. The second mouth was opened naturally on 1st August 2008 near "*Gabakunda*" located at about one km northward from the dredged mouth. During 2012 one more mouth was opened called as "*Dhalabali*" mouth by natural process. All these three mouths are playing crucial role to discharge away excess of rain and flood water along with silt and clay from the lagoon thereby maintaining the healthy status of the lagoon. These mouths are facilitating to increase the salinity level during post and premonsoon seasons (i.e. from November to June) by the ingress of more saline water coupled with evaporation from the surface. There are many Islands present in and around the lagoon premises. The noteworthy among them are Bhasra-Mundia, Ghantasila, Kalijugeswar, Malatikuda, Kankadakuda, Barunakuda, Chadheihagaa, Chadheiguhaa, Honeymoon Is., Badkuda, Sanakuda, Krushnaprasad Gargh, Parikuda, Somolonasi, Kalijai. All these islands are enriched with vegetation and also provide shelter to the many of the migratory birds and wild animals. The most important Island of Chilika is the Nalabana birds sanctuary. It appeared as the most suitable habitat for foraging, breeding and nesting ground of millions of migratory birds. This shallow region is also regarded as one of the important nursery ground for fishes and shellfishes. It further supports for colonizing of macrobenthic invertebrates (Mahapatro 2017).

3.1.3 Catchment Area

Chilika lagoon basin is strongly influenced with two catchment areas such as the (i) northern catchment area strongly influenced by the rivers like Daya, Bhargabi and Makara, the distributaries of river of Mahanadi and (ii) the western catchment which is strongly influenced by the rivers and rivulets like Luna, Ratnachira, Kani, Malaguni, Dhanua and Salia (Figure 3.4). Mahanadi delta contributed about 77 % fresh water discharges in to the lake, while the western catchment supports only 23 %. Similarly average annual sediment discharge is contributed by mainly by Mahanadi delta (75 %) and the rest (25 %) comes from the western catchment area. As reported by Mahapatro (2017) river Makara has maximum discharge of sediment into the lagoon.

3.1.4 Climate

Chilika lagoon climate is typically of tropical type. Largely influenced by southwest wind pattern and monsoon. The total rain fall over the area varies from 1190 to 1214 mm and mostly takes place during southwest monsoon period which accounted for nearly 80 % of the total rainfall. Again, the lagoon receives rainfall due to tropical cyclone and storm surges often occurred during the months of October and November and thunderstorms during March to May. The mean wind speed at Balugaon area is 0.678 kmph (kilometer per hour) and at Satapada it is 86.10 kmph (Mahapatro 2017). Wind speed with such a great velocity can churn the surface water and due to it slight increase of water level is marked during the premonsoon periods. Annually, three seasons are experienced in the lagoon area namely premonsoon (March to June), monsoon (July to October) and postmonsoon (November to February). Climatic features like temperature, wind direction and velocity, cloud coverage, precipitations humidity during these three seasons remains distinctly separate The solar days experienced over the lagoon area is around 255 days. The minimum solar radiation gained by the lagoon is 1.14 hours day^{-1}, while maximum of 10.13 hours day^{-1} is noticed (Mahapatro 2017).

Hydrography

The hydrographical parameter of Chilika lagoon is more dynamic and exhibit sharp spatiotemporal heterogeneity. Water temperature (^{0}C) in the lagoon is found higher in summer 35.5 ^{0}C (northern sector) and lowest of 18.9 ^{0}C (northern sector) in winter. Due to shallowness of the lake, the difference between surface and bottom water temperatures is found nearly 1 ^{0}C. Transparency of the lake water varied from 10 cm (Northern sector) in monsoon and 340 cm (Southern sector) in winter. The pH of the lagoon water varies from 6.6 (Northern sector) in monsoon to 9.3 (southern Sector) in winter. Salinity (ppt) varies from 0.0 (Northern sector) in monsoon to 35.6 in summer (outer channel). The dissolved oxygen content ranges from 2 mg l^{-1} (northern sector) in summer to 12.46 mg l^{-1} (northern sector) in winter. The difference between surface and bottom water DO is measured by about 1 mg l^{-1}. A well marked seasonal influence is observed among the parametrs like salinity, dissolved oxygen and water temperature (Mahapatro 2017).

Biodiversity

From biodiversity point of view, Chilika lagoon is very rich in floral and faunal diversity among all coastal lagoons of India. Ranging from microalgae to angiosperm and protozoa to mammals Chilika lake provides an excellent habitat to support to colonies the organisms from all spectrum of life. Since, four sector of Chilka Lake represent different kind of habitats, the biodiversity status is often found very high as compared other similar ecosystem in India and the world. Most of the species recorded from this ecosystem are alos described by IUCN as endangered, rare, threatened species etc.

Floral diversity

The floral components of Chilka lagoon mainly comprised of the micro-algae (Phytoplankton) and macro-algae (seaweeds), several forms of aquatic weeds, sea grasses and few species of mangroves. The Phytoplankton community is comprised of some autochthonous and alochthonous species of diatoms, representatives of dinoflagellates, blue green algae, green algae, and euglenoids. In total, 399 species of phytoplankton have been reported from these lake systems (Panda *et al.*, 2008). Among the phytoplankton, diatoms remained dominant as compared to other groups. Strong seasonal variation was marked in the abundance of phytoplankton. Some of the dominant diatoms are *Coscinodiscus gigas, Pinnularia alpine, Pleurosigma normanii, Navicula pygmaea,, Chaetoceros didymus, Asterionellopsis glacialis, Odontella sinensis, Biddulphia heteroceros.* Among dinoflagellates *Peridinium* and *Ceratium furca, Prorocentrum micans, Protoperidinium depressum* were found conspicuous where as among blue green alage *Anabena, Noctoc* and *Oscillatoria, Lyngbya aestuarii, Trichodesmium, Phormidium corium,* were observed as dominant throughout the years (Panigrahy *et al.*, 2009). A bimodal oscillation is observed in the annual cycle of phytoplankton density. Two peaks are observed corresponding March to May and in November every year. So far as the macrophyte vegetation is concerned; northern sector is dominated with fresh water macrophytes like *Phragmites karka, Eichornia crassipes, Hydrilla verticillata, Vallisneria spiralis, Ceratophyllum demersum, Chara, Naja* etc. The dominant macrophytes of central sector however, are comprised of the species of *Chara, Potamogetone* and *Schoenoplectus*. But, the southern sector is mostly dominated by sea grass such as *Halodule uninervis, Halodule pinifolia, Halophila ovalis, Halophila ovata, Halophila beccari* and *Ruppia maritima.* Rich sea grass vegetation is also seen in the Nalabana, Magarmukh, Krushnarasad Gadh, Rajhans, Arakhkuda, Bhabakundaleswar Is. regions. The most significant submerging type of macrophyte of Chilika lake is *Potamogetone pectinatus* which is found almost all the sectors of the lake area (Panda *et al.,* 2008).

Among the significant marine macroalgae *Enteromorpha intestinalis and Enteromorpha compressa* is seen in the islands of Kalijai, Pathara and Satpara region, *Ulva fasciata* and *Ulva lactuca* are predominant in Kalijai and Pathara segement. But, *Chaetomorpha linum* is distributed throughout the lake area. The red sea weed *Gracilaria verrucosa* is exclusively seen in Kalijai, Pathara, Honeymoon Island and Magarmukh area. The other macroalgae such as *Grateloupia filicina, Ceramium diaphanum and Polysiphonia sertularioides* are mostly confined to Kalijai, Pathara and Satpada area. In total 81 species including both micro and marcoalgae have been reported from the lake environment (Mahapatro 2017). The list of frequent occurring algal species in Chilika is given in the Table no. 1

The outer channel vegetation is all most same as in the southern sector. However a patch of mangroves is present in front of the dredged inlet and natural inlet at "Shipakuda" and "Gobakunda" region respectively. The common species is *Aegiceras corniculatum.* Besides these vegetations, the sand dunes of Nalabana, Kalijai, southern part Magarmukh and mud flats of outer channel area are endowed with some halophytes like *Salicornia brachiata.* Other sand dune vegetations include *Spinifex littoreus, Ipomea pescaprae, Macrotyloma ciliaum, Bulbostylis barbata* and *Hydrophylax maritama.* Besides these, some shoreline vegetations are also seen frequently among them *Casuarina equisetifolia* is conspicuous (Mahapatro 2017).

Faunal biodiversity

Chilika lake provides an excellent habitat that supports wide range of faunal diversity ranging from protozoans to mammalian groups. The faunal survey of Chilika lake was virtually started in the early parts of the 20th century by the researchers like Annandale, Kemp, Sewell, Preston, Chilton etc. They have given a detailed taxonomic description regarding, molluscs, polychaetes, amphipods, isopods, echinoderms etc. In the present context, the faunal community can be grouped as microbenthos, meio-benthos, macrobenthos, phytal benthos, zooplankton, fishes, shellfishes, birds, reptiles and mammals. The number of species corresponding to different phyla or group is given in Table-1. The Irrawaddy dolphin is regarded as the flagship species among the mammals of the Chilika lagoon which are mostly seen in the satapada, the outer channel areas and few region of central and southern side of the Chilika lagoon. The lagoon appeared as the most favoring ground for breeding and nesting of different organisms including birds and dolphins.

About 225 species of birds have been reported from the lagoon premises. Some of the dominant birds were Pelican, flamingos, Spoonbills, Egrets, Greylag geese, Storks and White Ibis, Shovellers and Pintails, Purple Herons, Brahminy Ducks Gray and Indian Roller, White bellied sea eagles. Most of the birds were congregated in the Nalabana bird Is., macrophyte vegetation areas, intertidal mud flats etc which considered as the most ideal destination for birds. These migratory birds find their destination from Caspian Sea, Baikal Lagoon and remote parts of Russia, Mongolia, Siberia, Iran, Iraq, Afghanistan and from the Himalayas to Chilika lagoon (Mahapatro 2017). The most common and migratory birds seen in Chilika are given in the Table no. 2

Out of a total of 406 species fishes contributed 321 species, 82 species of shell fishes (crab 53 sps. + prawn 29 sps.) along with three sp., of lobsters showing the rich diversity fish and shell fish diversity comprising the brackish water, marine and fresh water forms from Chilika lagoon (Table no. 3). Most of the fishes and shellfishes found in the lagoon are migratory in nature. There are 6 commercially important groups of fishes such as mullets, scianids, perches, belloniformis, threadfins and clupeids observed from the lake. Mullets are fish group represented by *Mugil cephalus, Liza macrolepis, Liza parsia, Liza subviridis* and *Liza tade.* Among Threadfins *Eleutheronema tetradactylum, Polydactylus sextarius* are conspicuous. Clupeiods like Nematalosa *nasus, Tenualosa ilisha, Hilsa kelee, Tenulosa hilsa* are important representatives. Among engruilidae, *Stolephorus indicus, Stolephorus commersonnii, Thryssa hamiltonii, Thryssa mystax* remain predominanat in this lagoon. The Perches such as *Lates calcarifer, Rhabdosargus sarba, Terapon jarbua, Sillago sihama, Datnioides polota* contribute visibly to total fish harvest. They are found in all sectors of Chilika lagoon. Among the Sciaenids, *Daysciaena albida, Dendrophysa russelii* where as *Strongylura strongylura* is the important species among the

Beloniformes found in the lagoon ecosystem. There are 53 species of crabs and 29 species of prawns and 3 species of lobster recorded from the lagoon (Table no. 3). Among the dominant crab species, mud crabs like *Scylla serrata* and *Scylla tranquebarica* are conspicous harvested from the lagoon ecosystem with a great economic value. Other crab species like *Portunus pelagicus, P. sanguinatus* are common and dominant in the lagoon. The shrimp population is dominated by *P. monodon, P. indicus, Metpeneaus dobsonii, M. monoceros.*, *Acetes* etc. Similalry, Stomatopod also contributed a significant proportion in the total catch. But, it doesn't have any market value so far the Chilika lake premices is concerend. Mostly, the central sector followed by outer channel area of the lagoon is appeared as the enriched region for exploitable fishery resources both from shellfish and fin fish point of view. The annual fisheries out from the lake are more than 10,000 tonnes per annum. The total list of fishes and shell fishes are given in the table no. 3. Table no. 4 represents the overall status of the biodiversity of Chilka lake.

Table 1: List of Common and rare Macro algae of Chilika lagoon

Sl. No.	Name of the species
1	*Cyanobacterium diachloros* Komárek & Anagnostidis
2	*Aphanocapsa marina* Hansgirg
3	*Merismopedia glauca* (Ehrenberg) Kützing
4	*Merismopedia punctata* Meyen
5	*Merismopedia warmingiana* Lagerheim
6	*Microcystis aeruginosa* Kützing
7	*Microcystis wesenbergii* (Komárek)
8	*Chroococcus limneticus* Lemmermann
9	*Pseudanabaena limnetica* Komárek
10	*Pseudanabaena minima* Anagnostidis
11	*Geitlerinema earlei* (Gardner) Anagnostidis
12	*Spirulina labyrinthiformis* Kützing ex Gomont
13	*Spirulina major* Kützing ex Gomont
14	*Spirulina subtilissima* Kützing
15	*Phormidium ambiguum* Gomont
16	*Oscillatoria limosa* Agardh exomont
17	*Oscillatoria perornata* Skuja
18	*Oscillatoria princeps* Vaucher ex Gomont

19	*Oscillatoria proteus* Skuja
20	*Oscillatoria sancta* Kützing ex Gomont
21	*Oscillatoria simplicis* sima Gomont
22	*Lyngbya aestuarii* Liebman ex Gomont
23	*Anabaena oscillarioides* Bharadwaj +
24	*Anabaena variabilis* Kützing ex Bornet et Flahault
25	*Ceramium diaphanum* Roth
26	*Gracilaria verucossa* (Huds.) Papenfuss
27	*Spirogyra* sp.
28	*Closterium venus* Kützing
29	*Chaetomorpha linum* (Mueller) Kützing
30	*Cosmarium awadhense* Prasad et Mehrotra
31	*Cosmarium decoratum* G.S. West
32	*Cosmarium lundellii* Delponte var. ellipticum W. et G.S. West
33	*Cosmarium miscellum* Skuja
34	*Cosmarium punctulatum* Brebisson
35	*Microspora willeana* Lagerheim
36	*Enteromorpha usneoides* (Bonnemaison) Agardh
37	*Enteromorpha compressa* (L.) Nees
38	*Enteromorpha intestinalis* (L.) Nees
39	*Euastrum dubium* Nägeli
40	*Pediastrum tetras* (Ehrenberg) Ralfs
41	*Chlorella protothecoides* Kützing
42	*Scenedesmus bijugatus* (Turpin) Kützing
43	*Scenedesmus calyptratus* Comas
44	*Scenedesmus dimorphus* (Turpin) Kützing
45	*Scenedesmus protuberans* Fritsch et Rich
46	*Uronema confervicolum* Lagerheim
47	*Uronema elongatum* Hodgetts
48	*Euglenophyta Euglena* acus Ehrenberg
49	*Euglena agilis* Carter
50	*Euglena caudate* Hübner
51	*Lepocinclis playfairiana* Deflandre
52	*Trachelomonas abrupt* Swirenko

53	*Trachelomonas hispida* Lemmermann
54	*Melosira decussate* (Ehrenberg) Kützing
55	*Odontella polymorpha* Kützing
56	*Coscinodiscus marginatus* Ehrenberg
57	*Chaetoceros decipiens* Cleve
58	*Cyclotella maxima* Kützing
59	*Cyclotella meneghiniana* Kützing
60	*Fragilaria crotonensis* Kitton
61	*Synedra crystalline* Kützing
62	*Synedra radians* Kützing
63	*Synedra tabulata* Kützing
64	*Synedra ulna* (Kützing) Hustedt
65	*Tabellaria flocculosa* (Roth)Kützing
66	*Cocconeis pediculus* Ehrenberg
67	*Pinnularia nodosa* (Ehrenberg) W. Smith
68	*Pinnularia subsimilis* Gandhi
69	*Pleurosigma javanicum* Grunow
70	*Pleurosigma normanii* Ralfs
71	*Navicula amphirhynchus* Ehrenberg
72	*Navicula major* Kützing
73	*Gomphonema micropus* Kützing
74	*Cymbella affinis* Kützing
75	*Amphora elliptica* Kützing
76	*Nitzschia acuta* Cleve
77	*Hantzschia amphioxys* Ehrenberg
78	*Epithemia gibberula* Grunow

Table 2: List of all 224 bird species seen in Chilika lagoon

Sl. No.	*Gnus-Species*
1	*Tachybaptus ruficollis* Pallas, 1764
2	*Podiceps cristatus* Linnaeus, 1758
3	*Pelecanus onocrotalus* Linnaeus, 1758
4	*Pelecanus philippensis* Gmelin, 1789
5	*Pelecanus crispus* Bruch, 1832

6	*Phalacrocorax niger* Vieillot, 1817
7	*Phalacrocorax fuscicollis* Stephens, 1826
8	*Phalacrocorax carbo* Linnaeus, 1758
9	*Anhinga melanogaster* Pennant, 1769
10	*Egretta garzetta* Linnaeus, 1766
11	*Egretta gularis* Bosc, 1792
12	*Ardea cinerea* Linnaeus, 1758
13	*Ardea purpurea* Linnaeus, 1766
14	*Casmerodius albus* Linnaeus, 1758
15	*Mesophoyx intermedia* Wagler, 1829
16	*Bubulcus ibis* Linnaeus, 1758
17	*Ardeola grayii* Sykes, 1832
18	*Butorides striatus* Linnaeus, 1758
19	*Nycticorax nycticorax* Linnaeus, 1758
20	*Ixobrychus sinensis* Gmelin, 1789
21	*Ixobrychus cinnamomeus* Gmelin, 1789
22	*Dupetor flavicollis* Latham, 1790
23	*Mycteria leucocephala* Pennant, 1769
24	*Anastomus oscitans* Boddaert, 1783
25	*Ciconia episcopus* Boddaert, 1783
26	*Ephippiorhynchus asiaticus* Latham, 1790
27	*Threskiornis melanocephalus* Latham, 1790
28	*Pseudibis papillosa* Temminck, 1824
29	*Platalea leucorodia* Linnaeus, 1758
30	*Phoenicopterus ruber* Linnaeus, 1758
31	*Phoenicopterus minor* Geoffroy Saint-Hilaire, 1798
32	*Dendrocygna bicolor* Vieillot, 1816
33	*Dendrocygna javanica* Horsfield, 1821
34	*Anser anser* Linnaeus, 1758
35	*Anser indicus* Latham, 1790
36	*Anser erythropus* Linnaeus, 1758
37	*Tadorna ferruginea* Pallas, 1764
38	*Tadorna tadorna* Linnaeus, 1758
39	*Sarkidiornis melanotos* Pennant, 1769

40	*Nettapus coromandelianus* Gmelin, 1789
41	*Anas strepera* Linnaeus, 1758
42	*Anas penelope* Linnaeus, 1758
43	*Anas poecilorhyncha* Forster, 1781
44	*Anas acuta* Linnaeus, 1758
45	*Anas clypeata* Linnaeus, 1758
46	*Anas querquedula* Linnaeus, 1758
47	*Anas crecca* Linnaeus, 1758
48	*Rhodonessa rufina* Pallas, 1773
49	*Aythya ferina* Linnaeus, 1758
50	*Aythya nyroca* Guldenstadt, 1770
51	*Aythya fuligula* Linnaeus, 1758
52	*Rallina fasciata* Raffles, 1822
53	*Gallirallus striatus* Linnaeus, 1766
54	*Rallus aquaticus* (Linnaeus, 1758)
55	*Amaurornis akool* Sykes, 1832
56	*Amaurornis phoenicurus* Pennant, 1769
57	*Porzana pusilla* Pallas, 1776
58	*Porzana fusca* Linnaeus, 1766
59	*Gallicrex cinerea* Gmelin, 1789
60	*Porphyrio porphyrio* Linnaeus, 1758
61	*Gallinula chloropus* Linnaeus, 1758
62	*Fulica atra* Linnaeus, 1758
63	*Hydrophasianus chirurgus* Scopoli, 1786
64	*Metopidius indicus* Latham, 1790
65	*Haematopus ostralegus* Linnaeus, 1758
66	*Rostratula benghalensis* Linnaeus, 1758
67	*Pluvialis fulva* Gmelin, 1789
68	*Pluvialis squatarola* Linnaeus, 1758
69	*Charadrius hiaticula* Linnaeus, 1758
70	*Charadrius dubius* Scopoli, 1786
71	*Charadrius alexandrines* Linnaeus, 1758
72	*Charadrius mongolus* Pallas, 1776
73	*Charadrius leschenaultii* Lesson, 1826

74	*Vanellus malabaricus* Boddaert, 1783
75	*Vanellus cinereus* Blyth, 1842
76	*Vanellus indicus* Boddaert, 1783
77	*Gallinago stenura* Bonaparte, 1830
78	*Gallinago gallinago* Linnaeus, 1758
79	*Lymnocryptes minimus* Brunnich, 1764
80	*Limosa limosa* Brunnich, 1764
81	*Limosa lapponica* Linnaeus, 1758
82	*Numenius phaeopus* Linnaeus, 1758
83	*Numenius arquata* Linnaeus, 1758
84	*Tringa erythropus* Pallas, 1764
85	*Tringa totanus* Linnaeus, 1758
86	*Tringa stagnatilis* Bechstein, 1803
87	*Tringa nebularia* Gunnerus, 1767
88	*Tringa ochropus* Linnaeus, 1758
89	*Tringa glareola* Linnaeus, 1758
90	*Xenus cinereus* Guldenstadt, 1775
91	*Actitis hypoleucos* Linnaeus, 1758
92	*Arenaria interpres* Linnaeus, 1758
93	*Limnodromus semipalmatus* Blyth, 1848
94	*Calidris tenuirostris* Horsfield, 1821
95	*Calidris canutus* Linnaeus, 1758
96	*Calidris alba* Pallas, 1764
97	*Eurynorhynchus pygmeus* Linnaeus, 1758
98	*Calidris minuta* Leisler, 1812
99	*Calidris ruficollis* Pallas, 1776
100	*Calidris temminckii* Leisler, 1812
101	*Calidris subminuta* Middendorff, 1853
102	*Calidris alpina* Linnaeus, 1758
103	*Calidris ferruginea* Pontoppidan 1763
104	*Limicola falcinellus* Pontoppidan, 1763
105	*Philomachus pugnax* Linnaeus, 1758
106	*Himantopus himantopus* Linnaeus, 1758
107	*Recurvirostra avosetta* Linnaeus, 1758

108	*Esacus recurvirostris* (Cuvier, 1829)
109	*Glareola pratincola* Linnaeus, 1766
110	*Glareola maldivarum* Forster, 1795
111	*Glareola lacteal* Temminck, 1820
112	*Larus heuglini* Bree, 1876
113	*Larus ichthyaetus* Pallas, 1773
114	*Larus brunnicephalus* Jerdon, 1840
115	*Larus ridibundus* Linnaeus, 1766
116	*Gelochelidon nilotica* J.E. Gmelin, 1789
117	*Sterna caspia* Pallas, 1770
118	*Sterna aurantia* Gray, 1831)
119	*Sterna bengalensis* Lesson, 1831
120	*Sterna bergii* Lichtenstein, 1823
121	*Sterna hirundo* Linnaeus, 1758
122	*Sterna albifrons* Pallas, 1764
123	*Sterna saundersi* Hume, 1877
124	Sterna acuticauda Gray, 1832
125	*Chlidonias hybridus* Pallas, 1811
126	*Chlidonias leucopterus* Temminck, 1815
127	*Chlidonias niger* Linnaeus, 1758
128	*Rynchops albicollis* Swainson, 1838
129	*Grus virgo* Linnaeus, 1758
130	*Elanus caeruleus* Desfontaines, 1789
131	*Milvus migrans* Boddaert, 1783
132	*Haliastur indus* Boddaert, 1783
133	*Haliaeetus leucogaster* Gmelin, 1788
134	*Haliaeetus leucoryphus* Pallas, 1771
135	*Circaetus gallicus* Gmelin, 1788
136	*Circus aeruginosus* Linnaeus, 1758
137	*Circus macrourus* (Gmelin, 1770)
138	*Circus melanoleucos* Pennant, 1769
139	*Accipiter badius* Gmelin, 1788
140	*Accipiter nisus* Linnaeus, 1758
141	*Pandion haliaetus* Linnaeus, 1758

142	*Falco tinnunculus* Linnaeus, 1758
143	*Falco chicquera* Daudin, 1800
144	*Falco subbuteo* Linnaeus, 1758
145	*Falco peregrinus* Tunstall, 1771
146	*Francolinus pondicerianus* Gmelin, 1789
147	*Perdicula asiatica* Latham, 1790
148	*Columba livia* Gmelin, 1789
149	*Streptopelia orientalis* Latham, 1790
150	*Streptopelia senegalensis* Linnaeus, 1766
151	*Streptopelia chinensis* Scopoli, 1786
152	*Streptopelia tranquebarica* Hermann, 1804
153	*Streptopelia decaocto* Frivaldszky, 1838
154	*Psittacula krameri* Scopoli, 1769
155	*Clamator jacobinus* Boddaert, 1783
156	*Hierococcyx varius* Vahl, 1797
157	*Eudynamys scolopacea* Linnaeus, 1758
158	*Centropus sinensis* Stephens, 1815
159	*Otus bakkamoena* Pennant, 1769
160	*Athene brama* Temminck, 1821
161	*Asio flammeus* Pontoppidan, 1763
162	*Cypsiurus balasiensis* Gray, 1829
163	*Apus affinis* Gray, 1830
164	*Alcedo atthis* Linnaeus, 1758
165	*Halcyon smyrnensis* Linnaeus, 1758
166	*Halcyon pileata* Boddaert, 1783
167	*Ceryle rudis* Linnaeus, 1758
168	*Merops orientalis* Latham, 1802
169	*Merops philippinus* Linnaeus, 1766
170	*Coracias benghalensis* Linnaeus, 1758
171	*Upupa epops* Linnaeus, 1758
172	*Megalaima zeylanica* Gmelin, 1788
173	*Megalaima haemacephala* Muller, 1776
174	*Mirafra cantillans* Blyth, 1844
175	*Mirafra erythroptera* Blyth, 1845

176	*Eremopterix grisea* Scopoli, 1786
177	*Calandrella cheleensis* Swinhoe, 1871
178	*Galerida cristata* Linnaeus, 1758
179	*Alauda gulgula* Franklin, 1831
180	*Hirundo rustica* Linnaeus, 1758
181	*Hirundo daurica* Linnaeus, 1771
182	*Motacilla maderaspatensis* Gmelin,1789
183	*Motacilla citreola* Pallas, 1776
184	*Motacilla flava* Linnaeus, 1758
185	*Motacilla cinerea* Tunstall, 1771
186	*Anthus richardi* Vieillot, 1818
187	*Anthus rufulus* Vieillot, 1818
188	*Anthus campestris* Linnaeus, 1758
189	*Pycnonotus jocosus* Linnaeus, 1758
190	*Pycnonotus cafer* Linnaeus, 1766
191	*Pycnonotus luteolus* Lesson, 1841
192	*Lanius cristatus* Linnaeus, 1758
193	*Monticola solitarius* Linnaeus, 1758
194	*Luscinia svecica* Linnaeus, 1758
195	*Copsychus saularis* Linnaeus, 1758
196	*Saxicoloides fulicata* Linnaeus, 1766
197	*Phoenicurus ochruros* Gmelin, 1774
198	*Saxicola caprata* Linnaeus, 1766
199	*Cercomela fusca* Blyth, 1851
200	*Chrysomma sinense* Gmelin, 1789
201	Turdoides malcolmi Sykes, 1832
202	*Turdoides striatus* Dumont, 1823
203	*Cisticola juncidis* Rafinesque, 1810
204	*Prinia inornata* Sykes, 1832
205	*Acrocephalus dumetorum* Blyth, 1849
206	*Acrocephalus stentoreus* Ehrenberg, 1833
207	*Phylloscopus trochiloides* Sundevall, 1837
208	*Sylvia curruca* Linnaeus, 1758
209	*Nectarinia zeylonica* Linnaeus, 1766

210	*Lonchura malacca* Linnaeus, 1766
211	*Passer domesticus* Linnaeus, 1758
212	*Ploceus manyar* Horsfield, 1821
213	*Ploceus philippinus* Linnaeus, 1766
214	*Sturnus roseus* Linnaeus, 1758
215	*Sturnus vulgaris* Linnaeus, 1758
216	*Sturnus contra* Linnaeus, 1758
217	*Acridotheres tristis* Linnaeus, 1766
218	*Acridotheres ginginianus* Latham, 1790
219	*Acridotheres fuscus* Wagler, 1827
220	*Oriolus oriolus* Linnaeus, 1758
221	*Dicrurus macrocercus* Vieillot, 1817
222	*Dendrocitta vagabunda* Latham, 1790
223	*Corvus splendens* Vieillot, 1817
224	*Corvus macrorhynchos* Wagler, 1827

Table 3: Fish and shellfish diversity of Chilika lagoon

Sl.no.	FISH Genus species
1	*Chiloscyllium indicum* Gmelin, 1789
2	*Carcharhinus leucas* Müller & Henle, 1839
3	*Carcharhinus limbatus* Müller & Henle, 1839
4	*Carcharhinus melanopterus* Quoy & Gaimard, 1824
5	*Glyphis gangeticus* Müller & Henle, 1839
6	*Scoliodon laticaudus* Müler & Henle, 1838
7	*Eusphyra blochii* Cuvier, 1816
8	*Sphyrna lewini* Griffith & Smith, 1834
9	*Pristis pectinata* Latham, 1794
10	*Rhynchobatus djeddensis* Forsskål, 1776
11	*Himantura imbricata* Bloch & Schneider, 1801
12	*Himantura marginata* Blyth, 1860
13	*Himantura uarnak* Forsskål, 1775
14	*Himantura walga* Müller & Henle, 1841
15	*Pastinachus sephen* Forsskål, 1775
16	*Aetobatus flagellum* Bloch & Schneider, 1801

17	*Aetobatus narinari* Euphrasen, 1790
18	Aetomylaeus nichofii Bloch & Schneider, 1801
19	*Notopterus chitala* Hamilton & Buchanan, 1822
20	*Currently, Chitala chitala* Hamilton, 1822
21	*Notopterus notopterus* Pallas, 1769
22	*Elops machnata* Forsskål, 1775
23	*Megalops cyprinoides* Broussonet, 1782
24	*Anguilla bengalensis bengalensis* Gray, 1831
25	*Anguilla bicolor bicolor* McClelland, 1844
26	Strophidon sathete (Hamilton, 1822)
27	*Lamnostoma orientalis* Mc Clelland, 1844
28	*Pisodonophis boro* Hamilton, 1822
29	*Pisodonophis Cancrivorus* Richardson, 1848
30	*Congresox talabonoides* Bleeker, 1853
31	*Muraenesox bagio* Hamilton, 1822
32	*Muraenesox cinereus* Forsskål, 1775
33	*Amblygaster leiogaster* Valenciennes, 1847
34	*Amblygaster sirm* Walbaum, 1792
35	*Anodontostoma chacunda* Hamilton, 1822
36	*Corica soborna* Hamilton, 1822
37	*Dussumieria acuta* Valenciennes, 1847
38	Dussumieria elopsoides Bleeker, 1849
39	*Ehirava fluviatilis* Deraniyagala, 1929
40	*Escualosa thoracata* Valenciennes, 1847
41	*Gonialosa manmina* Hamilton, 1822
42	*Gudusia chapra* Hamilton, 1822
43	*Hilsa kelee* Cuvier, 1829
44	*Nematalosa nasus* Bloch, 1795
45	*Sardinella fimbriata* Valenciennes, 1847
46	*Sardinella longiceps* Valenciennes, 1847
47	*Sardinella melanura* Cuvier, 1829
48	*Tenualosa ilisha* Hamilton, 1822
49	*Tenualosa toli* Valenciennes, 1847
50	*Setipinna phasa* Hamilton, 1822

51	Stolephorus baganensis Hardenberg, 1933
52	Stolephorus commersonnii Lacepède, 1803
53	*Stolephorus dubiosus* Wongratana, 1883
54	*Stolephorus indicus* Van Hasselt, 1823
55	*Thryssa gautamiensis* Babu Rao, 1971
56	*Thryssa hamiltonii* Gray, 1835
57	*Thryssa kammalensoides* Wongratana, 1883
58	*Thryssa malabarica* Bloch, 1795
59	*Thryssa mystax* Bloch & Schneider, 1801
60	*Thryssa polybranchialis* Wongratana, 1983
61	*Thryssa purava* Hamilton, 1822
62	Thryssa setirostris Broussonet, 1782
63	*Thryssa vitrirostris* Gilchrist & Thompson, 1908
64	*Chirocentrus dorab* Forsskål, 1775
65	*Ilisha elongata* Bennett, 1830
66	*Ilisha megaloptera* Swainson, 1839
67	*Ilisha melastoma* Bloch & Schneider, 1801
68	*Opisthopterus tardoore* Cuvier, 1829
69	*Chanos chanos* Forsskål, 1775
70	*Amblypharyngodon mola* Hamilton, 1822
71	*Chela cachius* Hamilton, 1822
72	*Laubuca laubuca* Hamilton, 1822
73	*Cirrhinus mrigala* Hamilton, 1822
74	*Cirrhinus reba* Hamilton, 1822
75	*Crossocheilus latius* Hamilton, 1822
76	*Danio rerio* Hamilton, 1822
77	*Esomus danricus* Hamilton, 1822
78	*Catla catla* Hamiltton, 1822
79	*Labeo boga* Hamilton, 1822
80	*Labeo calbasu* Hamilton, 1822
81	*Labeo gonius* Hamilton, 1822
82	*Labeo rohita* Hamilton, 1822
83	*Osteobrama cotio peninsularis* Silas, 1952
84	*Osteobrama vigorsii* Sykes, 1839

85	*Puntius chola* Hamilton, 1822
86	*Puntius sarana* Hamilton, 1822
87	*Puntius sophore* Hamilton, 1822
88	*Puntius ticto* Hamilton, 1822
89	*Puntius vittatus* Day, 1865
90	*Rasbora daniconius* Hamilton, 1822
91	*Rasbora rasbora* Hamilton, 1822
92	*Salmophasia bacaila* Hamilton, 1822
93	*Lepidocephalichthys guntea* Hamilton, 1822
94	*Mystus cavasius* Hamilton, 1822
95	*Mystus gulio* Hamilton, 1822
96	*Mystus vittatus* Bloch, 1794
97	*Sperata seenghala* Sykes, 1839
98	Siluridae (Eurasian catfishes)
99	*Ompok bimaculatus* Bloch, 1794
100	*Ompok pabda* Hamilton, 1822
101	*Wallago attu* Bloch & Schneider, 1801
102	*Ailia coila* Hamilton, 1822
103	*Eutropiichthys vacha* Hamilton, 1822
104	*Silonia silondia* Hamilton, 1822
105	*Pangasius pangasius* Hamilton, 1822
106	*Bagarius bagarius* Hamilton, 1822
107	*Bagarius yarrelli* Sykes, 1839
108	*Clarias magur* Hamilton, 1822
109	*Heteropneustes fossilis* Bloch, 1794
110	*Arius arius* Hamilton, 1822
111	*Arius maculatus* Thunberg, 1792
112	*Nemapteryx caelata* Valenciennes, 1840
113	*Osteogeneiosus militaris* Linnaeus, 1758
114	*Plicofollis argyropleuron* Valenciennes, 1840
115	*Plicofollis tenuispinis* Day, 1877
116	*Plotosus canius* Hamilton, 1822
117	*Plotosus lineatus* Thunberg, 1787
118	*Saurida tumbil* Bloch, 1795

119	*Trachinocephalus myops* Forster, 1801
120	*Liza macrolepis* Smith, 1846
121	*Liza melinoptera* Valenciennes, 1836
122	*Liza parsia* Hamilton, 1822
123	*Liza subviridis* Valenciennes, 1836
124	*Liza tade* Forsskål, 1775
125	*Liza vaigiensis* Quoy & Gaimard, 1825
126	*Mugil cephalus* Linnaeus, 1758
127	*Rhinomugil corsula* Hamilton, 1822
128	*Valamugil cunnesius* Valenciennes, 1836
129	*Valamugil seheli* Forsskål, 1775
130	*Valamugil speigleri* Bleeker, 1858-59
131	*Atherinomorus duodecimalis* Valenciennes, 1835
132	*Atherinomorus lacunosus* Forster, 1801
133	Strongylura leiura (Bleeker, 1850)
134	*Strongylura strongylura* Van Hasselt, 1823
135	*Xenentodon cancila* Hamilton, 1822
136	Hemiramphus far (Forsskål, 1775)
137	*Hyporhamphus limbatus* Valenciennes, 1847
138	*Oryzias dancena* Hamilon, 1822
139	*Aplocheilus panchax* Hamilton, 1822
140	*Hippocampus fuscus* Rüppell, 1838
141	*Ichthyocampus carce* Hamilton, 1822
142	*Hippichthys cyanospilos* Bleeker, 1854
143	*Ophisternon bengalense* Mc Clelland, 1844
144	*Macrognathus aral* Bloch & Schneider, 1801
145	*Macrognathus pancalus* Hamilton, 1822
146	*Mastacembelus armatus* Lacepède, 1800
147	*Pterois radiata* Cuvier, 1829
148	*Tetraroge niger* Cuvier, 1829
149	*Cociella crocodilus* Cuvier, 1829
150	*Kumococius rodericensis* Cuvier, 1829
151	*Platycephalus indicus* Linnaeus, 1758
152	Ambassidae (Perchlets, glass fishes)

153	*Ambassis ambassis* Lacepede, 1802
154	*Ambassis gymnocephalus* Lacepède, 1802
155	*Chanda nama* Hamilton, 1822
156	*Parambassis ranga* Hamilton, 1822
157	*Lates calcarifer* Bloch, 1790
158	*Epinephelus coioides* Hamilton, 1822
159	*Epinephelus lanceolatus* Bloch, 1790
160	*Epinephelus malabaricus* Bloch & Schneider, 1801
161	*Epinephelus tauvina* Forsskål, 1775
162	*Sillaginopsis panijus* Hamilton, 1822
163	*Sillago sihama* Forsskål, 1775
164	*Sillago vincenti* Mc Kay, 1880
165	*Lactarius lactarius* Bloch & Schneider, 1801
166	*Rachycentron canadum* Linnaeus, 1766
167	*Echeneis naucrates* Linnaeus, 1758
168	*Alectis indicus* Rüppell, 1830
169	*Alepes djedaba* Forsskål, 1775
170	*Atule mate* Cuvier, 1833
171	*Carangoides gymnostethus* Cuvier, 1833
172	*Carangoides praeustus* Bennett, 1830
173	*Caranx ignobilis* Forsskål, 1775
174	*Caranx melampygus* Cuvier, 1833
175	*Caranx sexfasciatus* Quoy & Gaimard, 1825
176	*Megalaspis cordyla* Linnaeus, 1758
177	*Parastromateus niger* Bloch, 1795
178	*Scomberoides commersonnianus* Lacepède, 1801
179	*Scomberoides lysan* Forsskål, 1775
180	*Scomberoides tala* Cuvier, 1832
181	*Scomberoides tol* Cuvier, 1832
182	*Selar boops* Cuvier, 1833
183	*Selar crumenophthalmus* Bloch, 1793
184	*Selaroides leptolepis* Cuvier, 1833
185	*Trachinotus blochii* Lacepède, 1801
186	*Trachinotus mookalee* Cuvier, 1832

187	Leiognathidae (Pony fishes)
188	*Eubleekeria splendens* Cuvier, 1829
189	*Gazza minuta* Bloch, 1795
190	*Leiognathus daura* Cuvier, 1829
191	*Leiognathus dussumieri* Valenciennes, 1835
192	*Leiognathus equulus* Forsskål, 1775
193	*Leiognathus fasciatus* Lacepède, 1803
194	*Nuchequula blochii* Valenciennes, 1835
195	*Nuchequula gerreoides* Bleeker, 1851
196	*Photopectoralis bindus* Valenciennes, 1835
197	*Secutor insidiator* Bloch, 1787
198	*Secutor ruconius* Hamilton, 1822
199	*Lutjanus argentimaculatus* Forsskål, 1775
200	*Lutjanus johnii* Bloch, 1792
201	*Lutjanus kasmira* Forsskål, 1775
202	*Lutjanus russellii* Bleeker, 1849
203	*Datnioides polota* Hamilton, 1822
204	*Gerres erythrourus* Bloch, 1791
205	*Gerres filamentosus* Cuvier, 1829
206	*Gerres limbatus* Cuvier, 1830
207	Gerres macracanthus Bleeker, 1854
208	*Gerres oyena* Forsskål, 1775
209	*Gerres phaiya* Iwatsuki & Hampstra, 2001
210	Haemulidae (Grunts & Rubberlips)
211	*Plectorhinchus nigrus* Cuvier, 1830
212	*Pomadasys argenteus* Forsskål, 1775
213	*Pomadasys kaakan* Cuvier, 1830
214	*Pomadasys multimaculatus* Playfair, 1867
215	*Acanthopagrus berda* Forsskål, 1775
216	*Acanthopagrus latus* Houttuyn, 1782
217	*Argyrops spinifer* Forsskål, 1775
218	*Crenidens crenidens* Forsskål, 1775
219	*Rhabdosargus sarba* Forsskål, 1775
220	*Nemipterus japonicus* Bloch, 1791

221	*Daysciaena albida* Cuvier, 1830
222	*Dendrophysa russelii* Cuvier, 1829
223	*Johnius carruta* Bloch, 1793
224	*Johnius dussumieri* Valenciennes, 1837
225	*Johnius belangerii* Cuvier, 1830
226	*Johnius coitor* Hamilton, 1822
227	*Johnius macropterus* Bleeker, 1853
228	*Nibea maculata* Bloch & Schneider, 1801
229	*Otolithes ruber* Bloch & Schneider, 1801
230	*Otolithoides biauritus* Cantor, 1849
231	*Otolithoides pama* Hamilton, 1822
232	*Paranibea semiluctuosa* Cuvier, 1830
233	*Protonibea diacanthus* Lacepède, 1802
234	Eleutheronema tetradactylum Shaw, 1804
235	*Leptomelanosoma indicum* Shaw, 1804
236	*Polydactylus plebeius* Broussonet, 1782
237	*Polydactylus sextarius* Bloch & Schneider, 1801
238	*Upeneus sulphureus* Cuvier, 1829
239	*Drepane punctata* Linnaeus, 1758
240	*Monodactylus argenteus* Linnaeus, 1758
241	*Nandus nandus* Hamilton, 1822
242	Terapontidae (Terapon perches)
243	*Pelates quadrilineatus* Bloch, 1790
244	*Terapon jarbua* Forsskål, 1775
245	*Terapon puta* Cuvier, 1829
246	*Terapon theraps* Cuvier, 1829
247	*Etroplus suratensis* Bloch, 1790
248	*Oreochromis mossambicus* Peters, 1852
249	*Ichthyoscopus lebeck* Bloch & Schneider, 1801
250	*Omobranchus zebra* Bleeker, 1868
251	*Butis butis* Hamilton, 1822
252	*Eleotris fusca* Forster, 1801
253	*Eleotris melanosoma* Bleeker, 1852
254	*Acentrogobius cyanomos* Bleeker, 1849

255	*Acentrogobius griseus* Day, 1876
256	*Acentrogobius masoni* Day, 1873
257	*Acentrogobius viridipunctatus* Valenciennes, 1837
258	*Amoya madraspatensis* Day, 1868
259	*Bathygobius fuscus* Ruppell, 1830
260	*Bathygobius ostreicola* Chaudhuri, 1916
261	*Brachygobius nunus* Hamilton, 1822
262	*Drombus globiceps* Hora, 1923
263	*Glossogobius giuris* Hamilton, 1822
264	*Glossogobius mas* Hora, 1923
265	*Gobiopterus chuno* Hamilton, 1822
266	*Oligolepis acutipennis* Valenciennes, 1837
267	*Oligolepis cylindriceps* Hora, 1923
268	*Oxyurichthys microlepis* Bleeker, 1849
269	*Oxyurichthys tentacularis* Vallenciennes, 1837
270	Parapocryptes rictuosus Valenciennes, 1837
271	Periophthalmus kalolo Lesson, 1831
272	*Psammogobius biocellatus* Valenciennes, 1837
273	*Pseudapocryptes elongatus* Cuvier,1816
274	Pseudogobius javanicus Bleeker, 1856
275	*Stigmatogobius minima* Hora, 1923
276	*Taenioides buchanani* Day, 1873
277	*Trypauchen vagina* Bloch & Schneider, 1801
278	*Yongeichthys criniger* Valenciennes, 1837
279	*Ephippus orbis* Bloch 1787
280	*Platax orbicularies* Forsskål, 1775
281	*Scatophagus argus* Linnaeus, 1766
282	Siganidae (Spinsfoots, Rabbitfishes)
283	*Siganus canaliculatus* Park, 1797
284	*Siganus javus* Linnaeus, 1766
285	*Siganus vermiculatus* Valenciennes, 1835
286	*Acanthurus mata* Cuvier, 1829
287	*Sphyraena jello* Cuvier, 1829
288	*Sphyraena putnamae* Jordan & Seale, 1905

289	*Eupleurogrammus glossodon* Bleeker, 1860
290	*Trichiurus lepturus* Linnaeus, 1758
291	*Lepturacanthus savala* Cuvier, 1829
292	*Euthynnus affinis* Cantor, 1849
293	*Rastrelliger kanagurta* Cuvier, 1816
294	*Scomberomorus lineolatus* Cuvier, 1829
295	*Anabas cobojius* Hamilton, 1822
296	*Anabas testudineus* Bloch, 1792
297	Colisa fasciata (Bloch & Schneider, 1801)
298	*Colisa lalia* Hamilton, 1822
299	*Channa gachua* Hamilton, 1822
300	*Channa marulia* Hamilton, 1822
301	*Channa punctata* Bloch, 1793
302	*Channa striata* Bloch, 1793
303	Paralichthyidae (Lefteye flounders)
304	*Pseudorhombus arsius* Hamilton, 1822
305	*Pseudorhombus micrognathus* Norman, 1927
306	Pseudorhombus triocellatus (Bloch & Schneider, 1801)
307	*Brachirus orientalis* Bloch & Schneider, 1801
308	*Solea ovata* Richardson, 1846
309	*Cynoglossus lida* Bleeker, 1851
310	*Cynoglossus lingua* Hamilton, 1822
311	*Cynoglossus puncticeps* Richardson, 1846
312	*Triacanthus biaculeatus* Bloch, 1786
313	*Abalistes stellaris* Bloch & Schneider, 1801
314	*Arothron reticularis* Bloch & Schneider, 1801
315	*Arothron stellatus* Bloch & Schneider, 1801
316	*Chelonodon patoca* Hamilton, 1822
317	*Lagocephalus lunaris* Bloch & Schneider, 1801
318	*Takifugu oblongus* Bloch, 1786
319	*Tetraodon cutcutia* Hamilton, 1822
320	*Tetraodon fluviatilis* Hamilton, 1822
321	*Diodon hystrix* Linnaeus, 1758

Crab	
322	*Benthopanope indica* de Man, 1887
323	*Calappa philargius** Linneaus, 1758
324	*Camptandrium sexdentatum* Stimpson, 1858
325	*Cardiosoma carnifex* Herbst, 1796
326	*Charybdis callianasa* Herbst, 1801
327	*Charybdis cruciatai* Herbst, 1794
328	*Charybdis hellerii**Edwards, 1867
329	*Diogenes custos** Fabricious, 1798
330	*Doclea hybrid* H. Milne Edwards, 1834
331	*Dotilla intermedia* de Man, 1888
332	*Dotilla myctiroides* H. Milnne Edwards, 1852
333	*Dotilla pertinax* Kemp, 1915
334	*Ebalia malefactrix* Kemp, 1915
335	*Elamina (Trigonoplax) cimex* Kemp, 1915
336	*Emerita emeritus* Latreille, 1804
337	*Gecarcoidae lalandii** Edwards, 1837
338	*Heteropanope indica* de Man, 1887
339	*Hippa pacifica* Dana 1852
340	*Leipocten sardidulum* Kemp, 1915
341	*Leucosia anatum* Herbst, 1783
342	*Leucosia pubescens** Miers, 1877
343	*Macrophalmus gastrode* Kemp, 1915
344	*Matuta lunaris* Forskal, 1775
345	*Matuta planipes** Fabricius, 1798
346	*Matuta victor* Fabricius, 1781
347	*Metopograpsus messor* Forskal, 1775
348	*Neosarmatium meinerti** de Man, 1887
349	*Ocypoda ceratophalma* Pallas, 1772
350	*Ocypoda macrocera* H. Milne Edwards, 1852
351	*Ocypoda platytaris* H. Milne Edwards, 1852
352	*Ocypode cordimana** Desmarest,1825
353	*Pachygrapsus propinquus* De Man, 1908
354	*Parthenopus echinatus** Herbst, 1790

355	*Persephona lichtensteinii* * Leach, 1817
356	Philyra alcocki Kemp, 1915
357	*Philyra malefactrix* Kemp, 1915
358	*Philyra scabriuscula** Fabricius, 1798
359	*Plagusia depressa tuberculata* Lamarck, 1818
360	*Podophthalmus vigil* Fabricius, 1798
361	*Portunus pelagicus* Linnaeus, 1758
362	*Portunus sanguinolentus* Herbst, 1783
363	*Ptychognathus onyx* Alock, 1900
364	*Schizophrys aspera** Edwards, 1834
365	*Scylla serrata* Forskal, 1775
366	*Scylla tranqubarica* Fabricius, 1798
367	*Sesarma batavicum* Moreira, 1903
368	*Sesarma plicatum* Latreille, 1803
369	*Sesarma quadrata* De Haan, 1850
370	*Sesarma tetragonum* Alcock, 1900
371	*Thalamita crenata* Ruppel, 1830
372	*Uca annulipes* H. Milne Edwards, 1837
373	*Uca vocans* Linnaeus, 1758
374	*Varuna litterata* Fabricius, 1798
Prawn	
375	*Penaeus monodon* Fabricius, 1798
376	*Penaeus indicus* H. Milne Edwards, 1837
377	*Penaeus canaliculatus* Oliver, 1811
378	*Penaeus semisulcatus* De Haan, 1844
379	*Metapenaeus monoceros* Fabricius, 1798
380	*Metapenaeus affinis* H. Milne Edwards, 1837
381	*Metapenaeus dobsoni* Miers, 1878
382	*Metapenaeus ensis* De Haan, 1844
383	*Lucifer hanseni* Nobili, 1905
384	*Pontophilus hendersoni* Kemp, 1915
385	*Macrobrachium lamarrei lamarrei* H. Milne Edwards, 1837
386	*Macrobrachium malcolmsoni* H. Milne Edwards, 1837
387	*Macrobrachium rude* Heller, 1862

388	*Macrobrachium scabriculum* Heller, 1862
389	*Exopalaemon styliferus* H. Milne Edwards, 1840
390	*Periclimenes (Periclimenes) indicus* Kemp, 1915
391	*Periclimenes (Harpilius) demani* Kemp, 1915
392	*Macrobrachium rosenbergiii* De Man, 1879
393	*Macrobrachium equidens* Dana, 1852
394	*Ogyrides striaticauda* Kemp, 1915
395	*Athanas polymorphus* Kemp, 1915
396	*Alpheus crassimanusi* Heller, 1865
397	*Alpheus malabaricusi* Fabricius, 1775
398	*Alpheus paludicola* Kemp, 1915
399	*Caridina nilotica* P. Roux, 1833
400	*Caridina propinqua* De Man, 1908
401	*Leptochela aculeocaudata* Paulson, 1875
402	*Callianassa (Callichirus) maxima* A. Milne Edwards, 1870
403	*Upogebia (Upogebia) heterocheir* Kemp, 1915
Lobster	
404	*Panulirus polyphagus* Herbst, 1793
405	*Panulirus ornatus* Fabricius, 1798
406	*Thenus orientalis* Lund, 1793

3.1.9 Environmental issues of the Lake

Considering the ecosystem services and functioning rendered by Chilika lagoon, it was designated as one of the Ramsar Sites from India in 1981. Because of its it vast biodiversity which is reflected in the Table no. 4 that represents the overall biodiversity of the lake. The lake gets major chunk of internation attaintion. However, the lagoon environment had experienced many visible changes over the years under the influence of a series of natural perturbations and anthropological activities. The important environmental degradations the lake faced in recent years are siltation, salinity reduction, eutrophication, macrophyte infestation, over fishing and chemical pollution including oil & pesticide pollution. The detail account of environmental degradations and their influences on the biota of the Lake has been discussed earlier by many researchers (Panigrahy, 2000; Panda *et al.,* 2008; Sahu *et al.,* 2014). In all these publications more investigations on each aspect of the lagoon has been emphasized. The present study is under taken to elucidate the changes that have occurred as regards to the macro-zoobenthic community of the lake basin after the hydrological interventions.Table no. 4 represents the overall status of the biodiversity of Chilka lake.

Table 4: Represents the overall biodiversity of Chilka lake.

Faunal group	No. of Species
Floral component	
Phytoplankton	399
Micro and macro sea algae	81 (Mohanty et al., 2013)
Plants	720
Sea grass	5
Sea weed	14 (Sahoo et al., 2005)
Faunal group	
Zooplankton	170
Protozoa	61
Porifera	7
Coelenterates	7
Nematodes	37
Mollusca	136
Annelids	31
Crustacea	28
Decapoda	34
Echinodermata	5
Protochordata	1
Fish	324
Crab	53
Prawn	29
Amphibians	7
Lobster	3
Reptile	30
Birds	225
Mammals	18
The list may be increased	

Ecosensitive regions of Chilika

Nalabana:

It is the major winering destination of migratory birds in Asia. Geographically located at the centre of the (Figure-6) thus, fall in the central sector covering an area of 15.53 km^2 (6.00 sq mi). This is a well designated bird's destination in India and abroad, delacred as a sandtury in 1987 under Wildlife Protection Act (1973). It is a temporary Is. which is completely immersed in the lake water during monsoon. This shallow Island gradually exposed as a semi dry-semi wet Island endowed with many of the macro algae and sea grass vegetation during winter and completely exposure as an Island in the summer. The temporary exposure of this shallow Is. shows the growth of an array of bottom dwelling flora and fauna and becomes one othe nutrient rich ecosystem for migratory birds, juveniles of fishes, prawn and crabs etc.

Figure 6: *Nalabana Bird Is. inside the map of Chilika lake, Bay of Bengal.*

***Figure 7:** Image showing Migratory Bird congregation at Nalban Is.*

Towards the end of monsoon the Nalabana Is. is covered with large no. of fresh water, brackish water weed. Later on the marine weeds replace the fresh water weeds as the salinity starts to increase gradually. This dynamism in the salinity directly controls the vegetation as well as benthic invertebrate community. Thus, the long legged birds first arrive in the sanctuary and when the water level recceds gradually short legged birds prefers to come. Some the birds seen in Nalabana Bird sanctuary are those of long legged waders, lesser flamingos, Goliath heron, grey herons, purple herons, egrets, spoonbills, storks and black-headed ibis. Some of the rare birds were alos rcorded such as Asiatic dowitchers (NT), Dalmatian pelican (VU), Pallas's fish-eagles (VU), the very rare migrant spoon-billed sandpiper (CR) and spot-billed pelican (NT). The white-bellied sea eagle, pariah kite, brahminy kites, kestrel, marsh harriers are common to see. Among the short-legged shorebirds include plovers,collaredpratincole, ruff, dunlin, snipes and sandpipers. Mudflat ecosystem contains Larks, wagtails and lapwings while longer-legged avocets, stilts and godwits can bee seen to feed in the deeper water. The densly vegetated areas often showed support moorhens, coots and jacanas where as pond herons and night herons arecommon in the shorelines alongwith kingfishers and rollers, little cormorants, perches are seen around the lakeassociated with brahminy ducks, shovellers, pintails, gadwall, teals, pochards, geese and coots are the some of the conspicuous species to notice (Figure-7). The details of the list of the birds are given in table no. 2.

Kalijai Island:

The Kalijai Is land is a three kilometer long elongated Is. located between 85°13'1.02"E & 19°39'53.66"N to 19°39'56.55"N and 85°12'43.16"E (Figure-8). It falls under the southern sector of Chilika lake. It is a popular piligrimage place note only in Odisha but alos in India. The basement of this Is. made up of rock and the soreline is alos the rockey. Probably it is an extended part of the easternghat moutaning range and detached from the mainlad due to the some natural reasosn. According to a poem written by a great Odia poet Pt. Godavarisha Mishra the Jaai is a girl who was marrying a person to Parikuda Is. while going to her mother-in-law her boat was capsized in a severe cyclonic storm and Kaalijai met a tragic death. Soonafter her death to provide obituary in her memory, a temple is built by the king of Banasur in the year 1717. The rockeyshore of kalijai Is. is the rich resource of a variety of green algae and red algae namely *Graccilaria verucosa*, *Gracillaria lichenoides*, *Ulva lactuca*, *Ulva* sp., *Enteromorpha intestinalis*, *Entermorpha compressa*, *Chaetomorpha linum* etc. According to a study Kalijai Is is the larest repository of marine algae with respect to species number, abundance and biomass. This region is found much richer in the sessile benthic macroinvertebrate community. The faunal groups those have made their ultimate destination this locality are those of bivalves, gastropods, isopods, amphipods, and many of the biofouling organisms. The most dominanat molluscan species found in this region is *Modiolus striatulus* and *Nassarius stolatus*.

Figure 8: *Marine algae and molluscan bivalve shells attached around the rockey shore of Kalijai Is.*

Bhaba-Kundalwswar Temple:

At the very close vicinity of the outer channel area of Chilika lake, Bahbakundaleswar temple is located with its antique architechture and beauty. It is the temple of Lord Shiva (Figure-9). Historians believe that this temple is made by the powerful king of then Kalinga/Utkal Maharaja Purrosttam Deva during his Kaanchi abhijaan. According to the local sources, this temple is built about 800 years ago and placed under the sand because of the natural deposition of beach sand for a longer period of time. The excavation that is going on to rejuvenate the previous glory of the temple can be noticed nicely. The design of this temple is quite similar to that Sri Jagnnath Temple of Puri and Surya temple of Konark. The architecture is quite similar to these temples. Thus, it is presumed that the temple must be constructed more than 800 years ago. The outer sculpture of the temple is marvelous which can be seen by the photographs attached. The main temple seems much similar to Jagannath temple but the frontal design of Jagmohan is similar to that of Konark temple. This is amazing to see the combination design of Puri temple and Konark temple at one place. It is also said that there was a lamp over this temple that acted as a light house for the sea sailors of Odisha. This diligent construction is our pride and glory. Steps should be taken to protect this ancient structure from marine corrosion. It can be a great tourist spot since the temple located at the very close vicinity of Chilika and Bay of Bengal. Thus hope that it can find a place in the map of Incredible India as another pride of Odisha. The nearby intertidal region of Bhabkundaleswar temple showed very rich bed of molluscan community and few mangrove vegetations. Species like *Crasssostra cuttacknisis*, *Saccostrea cucculata*, *Anomia* sp., *Telescopium telescopium*, *Cerithidea* sp., etc are profoundly observed in its bed which can be easily visulaised during low tide periods. This region is treasure for the benthoic macrofunal communities, fishes, mud skipeers etc.

Figure 9: *BhabKundaleswar Temple of Outer channel area of Chilika lake . Bottom three images showing the oyster shell bank, window pane oyster and mangrove vegetation respectively.*

3

Collection, Preservation, Identification & Classification of Mollusca

Molluscs are generally described as soft-bodied animals, a majority of which are covered by a hard calcareous shell. The shell may consist of one, two or many pieces which may be internal and cartilaginous. This includes a heterogeneous group under different names viz. snails, slugs, mussels, oysters, clams, cuttlefish, squids, octopuses, etc. From the Cambrian period the mollusc have been appeared in the earth crust and distinguished into seven different classes. Out of which five classes are exclusively found in India. Molluscan diversity in Chilika is very rich. The 32 km of outer channel area is rich treasure of marine molluscan fauna while, the central and southern sector of Chilika becomes the most suitable habitat of marine and brackish water species of mollusc. The fresh water dominant species are found in the Northern sector. In this chapter we shall discuss regarding the Collection, preservation, and identification & classification of molluscan fauna of Chilika lagoon.

Collection:

During April 2007 to March 2010, the entire outer channel area and Chilika main lake premises was scrutinised for the collection of molluscan communities (Figure-10 A). Sampling was made on seasonal basis from 24 prefixed stations (Figure-10 B). The sediment samples were collected by the helps of a van veen type of grab sampler (Figure -11) having the surface area of (250 cm^2) in the deeper areas while the intertidal sampling was made during low tide periods (Figure-12). Molluscan samples comprised of Gastropods and bivalves were collected by hand picking or by using a drag net from the intertidal areas of the outer channel. Dead shells and the shell carried out by the ghost crab/hermit crab were also obtained for observation.

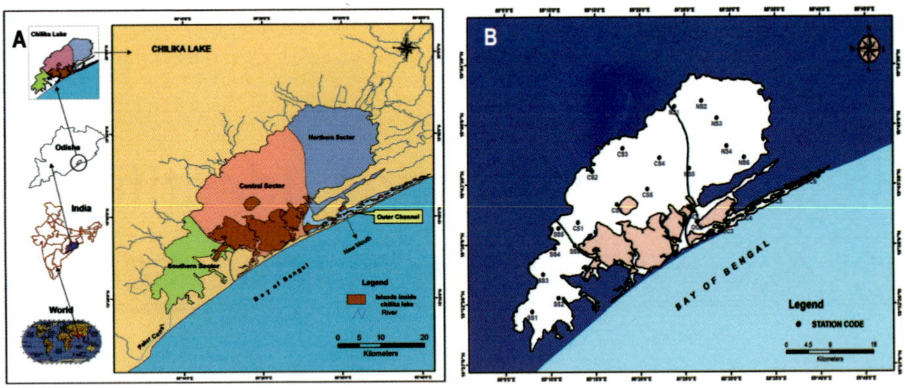

Figure 10: *Location of Chilika lake (A) and Sampling stations of Chilika lake (B).*

Figure 11: *Sampling and data collecting photographs of Mollusca from Chilika lake.*

Figure 12: *Molluscan shell bank found at the low tide periods of outer channel area in the Chilika lake.*

Preservation:

The collected live sample were allowed narcotised first by keeping the fresh water and air tightened. Then their live tissue is removed out. The shells whose muscle tissue parts were not able to remove out were preserved in the 10 % formalin solution. Then the dead shells of bivalves and gastropods were allowed to settle in diluted hydrochloric acid solution to remove out the outer debris found on mollusc so that the natural colour of the mollusc is exposed out. Then after, all the specimens were preserved in 90% ethyl alcohol the reason is to narcotise the specimen first before preserving in to the alcohol. When all the samples were narcotised. They are transferred to 5 % neutralised formaldehyde solution. After the preservation process over the identification of the collected shells were done by adopting standard identification literature, published keys found elsewhere and by following online available identification and taxonomic classification such as WoRMS, EOL, GBIF, NIO biosearch, and Marine Species Identification Portal (MSIP). A lot of literatures, handbook, and monographs published by Zoological Survey of India, NIO-GOA and CMFRI-ICAR were also followed. A sincere approach was adopted to identify all the collected specimens up to species level. Collection of information and corroboration of data such as diagnostic features, habitat distribution and photography including the date of collection, name of the collector, location etc were done by the first author. There are certain ethics for the collection and preservation of the shells are being followed throughout the world. This is considered as the prime concern of all naturalists and biologists who worked up on these animal communities. These codes are:

- *Cause minimal disturbance to its habitat*: Habitat destruction is the most potent threat to al1 biota including molluscs.
- *Observe the laws regulating access sites*: Protected Areas, (viz., Biosphere Reserves, National Parks, Sanctuaries etc.) the rules and laws have been established by the Government to protect habitats and their constituent communities.
- In the areas where, observations, collections are permitted, disturbance to the habitats (vegetation, boulders) should be minimal, so that the other communities living in the system are not disturbed.
- Collect minimum number of specimens' necessary, i.e., indiscriminate collection has devastating effect on the local population.

Following points should be taken in to consideration while collecting the molluscan samples.

- Local, national and international laws governing species and habitats should be abided
- Species in the reproductive stages (such as during egg laying, breeding and other developing stages) should not be collected
- Juvenile, live individuals with damaged shells are of little value for the collector and therefore, collection of such specimens should be avoided
- Only required number of specimens that are strictly for the study should be collected.
- Empty and shells washed ashore are to be preferred
- Photography in the natural conditions is mostly preferred for the naturalists

During the present study all the above mentioned points were taken in to the consideration.

Classification of Mollusc:

Mollusca undergo some simpler body structure to the complex with respect to their habit and habitat. Albeit, general structural design remains identical. For example, Aplacophora or Solenogasters are worm like, with distinguished head and the body with calcareous spicules. Similarly in case of Polyplacophora or chitons distinct head observed whereas eyes, tentacles absent along with eight longitudinal valves that represent the shell. But, Monoplacophora resemble the gastropod limpets that differ in the metameric repetition of many parts. The gastropods or snails are regarded as typical mollusc with well developed head and spirally coiled shell. The bivalve also called as Pelecypods don't posses head, radula and are entirely enveloped with in a lateral valves shell. The most different structures are said to be found in the tusk shells or scaphopoda. These groups do not posses gills but roughly differentiated with head which is enclosed with in a in a tusk-shaped shell with apical lobe (Ramakrishna and Dey 2003).

However, the largely diverse body structure is found among Cephalopoda, which includes Nautilus, Octopus and squids. The most amazing features of this groups are found is the foot is modified in to a circle of arms endowed with the suckers. They are the some large bodied invertebrate in the marine condition.

Because of the diverse body structure, food and feeding habit, it is very difficult to classify the mollusc under a single type consideration. However, many changes were made in the recent past to upgrade the taxonomic classification with suitable justifications. In recent times the classification of Vaught (1988), is considered as the ideal one who classified the Phylum Mollusca Cuvier, 1795 in to 7 different classes namely Class 1. Aplacophora von Jhering, 1876, Class 2. Polyplacophora de Blainville, 1816, Class 3, Monoplacophora Wenz, 1940,Class 4. Gastropoda Cuvier, 1795, Class 5. Bivalvia Linnaeus, 1758, Class 6. Scaphopoda Bronn, 1862, Class 7., Cephalopoda Cuvier, 1795 (Ramakrishna and Dey 2003). Five out of seven classes are found in Indian sub-continental region namely, Cephalopoda, Scaphopoda, Bivalvia, Gastropoda, Polyplacophora. A total of 242 numbers of molluscan familes were recorded from marine habitat from which gastropod is the largest number of families (140 no.), Bivalves having 61 number of families, scaphopoda 3 species, Cephalopoda 28 families and polyplacophora having 10 families. Fresh water and terrestrial gastropod shared 14 and 26 families. Land bivalves are reported till date while fresh water bivalves are represented by 8 numbers only. The details of the seven classes of mollusc are given as follows.

Class 1. Aplacophora

Body is vermiform, bilaterally symmetrical, untorted molluscs without head, mantle, foot, shell or nephridia; cuticle with calcareous spicules; digestive tract straight with a radula; a pair of coelomoducts in the form of gonoducts, which open into the terminal part of the intestine or directly. Species are exclusively of marine occurring from shallow regions to the deep waters i.e. from 18-4000 m found among the hydroid colonise to which they feed up on. Commonly distributed in Antarctica, European coasts, Atlantic coast of North America Mediterranean, Indo-Pacific (few representation), North Pacific.

Class 2 POLYPLACOPHORA

The class Polyplacophora are characterised as dorso-ventrally flattened, bilateral symmetrical, untorted body, head do not have of eyes and tentacles, mantle, and external gills and radula, shell comprised of eight, longitudinal series of successive pieces, foot is flat bordered by lateral grooves containing gills. Coiled intestine, terminal anus a pair of metanephidia connected to the pericardium and a pair gonoducts. They are commonly known as the chitons or coat of mail shells. Cosmopolitan in nature occur in all seas, frequently found in the rocky shores in the intertidal or upper littoral region. In India, 32 species of Chitons are observed by which are found occurring on the coasts of Andaman and Nicobar Islands, Gulf of Mannar, Coast of Andhra, and in Gujarat coast (Ramakrishna and Dey 2003).

Class 3. MONOPLACOPHORA

This class is the bilateral symmetrical univalve shells segmented unsorted body with head, foot, mantle and radula. External gills, nephredia, muscles, nerve branches, with terminal anus. This is represented by a single genus and four species mostly found in the deep waters of Atlantic Sea on dark, muddy clay, ranged at the depth of 2700 m to 5700 m(Ramakrishna and Dey 2003).

Class 4. GASTROPODA

The body is unsegmented, asymmetrical, spirally coiled univalve shell, well developed head bearing eyes, tentacles, and well developed foot having ventral surface is flat forming a creeping sole, mantle lining the last whorl of the shell, radula well developed, visceral mass spirally coiled, with paired or unpaired heart auricle, osphridium gill and nephridium, with a single reproductive system. Sexes are separate or united, oviparous. It comprised of about 80% of the known molluscan species in the world. Distribution is ranged from intertidal zone to abyssal zone near about 8200 metre. Five subclasses are further categorised under this clases such as Pulmonata, Gymnomorpha, Heterobranchia, Opisthobranchia and Prosobranchia. The economically important families in Gastropoda are those of Haliotidae, Olividae, Patellidae, Trochidae, Turbinidae, Strombidae, Potamididae, Muricidae, Mitridae, Vasidae, Turbinellidae, Conidae and Cassidae(Ramakrishna and Dey 2003).

Class 5. BIVALVIA

Bilaterally symmetrical, laterally compressed, untorted, enclosed in a calcified shell of two valves, hinged together mid dorsally. Outer surface of the shell bears longitudinal or concentric growth striations. Ventral surface of each valve of the shell bears impressions of ligaments,which opens the shell and adductor muscles that close it. A ventral axe shaped or wedges shaped foot present. The mantle is bilobed, without head pharynx, jaws, radula or tentacles, with two auricles, a pair of gill, nephridia and gonads, dioecious or hermaphrodite are seen. The members are mostly of marine or freshwater in nature(Ramakrishna and Dey 2003). It has five sub classes namely Protobranchia, Pteriomorpha, Paleoheterodonta, Heterodonta and Anomalodesmata.

Class 6. SCAPHOPODA

Symmetrically, without eyes, tentacles, gills, enclosed in a tusk like shell open at both ends, mantle tubular through fusion of its edges, foot enlarged, mouth surrounded by lobular outgrowths, base of snout with pair clusters of protrusible filaments, rudimentary heart, diocious. (Ramakrishna and Dey 2003).

Class 7. CEPHALOPODA

Body is external bilateral symmetry, with spiral chambered shell otherwise usually with or without reduced shell imbedded in the dorsal mantle, highly cephalised, with large eyes, jaw and radula, foot modified into arms encircling the mouth and beset with suckers (except Nautiloids), 2 – 4 bipectinate gills, heart 2- 4 chambered, closed circulatory system, nephridia four, dioecious. This class is divided into two subclasses namely Nautiloidea and Coleoidea (Ramakrishna and Dey 2003).

Identification of Bivalves:

It is the second largest group after gastropod among the phylum mollusc. These are bearing of a different body structure when compared with the gastropods since, the muscular body parts of bivalve are completly enclosed in the two valves. Being compressed laterally, bivalves often possesses a sedentary mode of life or it can move a very little by the help of its muscular foot. The shells of the bivalves can be divided in two equal halves, also called as valves. They are hinged together dorsally by an elastic, chitinous, external or internal ligament. A set of interlocking teeth prevents the valves from sliding along each other. This is highly helpful during predation, reproduction, locomotion and escape our from predators. A pair of adductor muscles is responsible to shut the valves. It has been also marked that the head in bivalves is reduced, most likely because of its sedentary lifestyle. Even radula, eyes or tentacles is lost among the bivalves. Although, few of them have these organs. Some of the bivalves along the mantle margin may also possess secondary tentacles and eyes. Mouth of the bivalve is located well inside the animal, and a pair of fleshy labial palps helps direct the food particles toward the mouth after these particles have been collected and sorted by the ctenidia. They can be deposit-feeders (subclass Protobranchia). To collect food particles from the bottom surface, these bivalves can utilize their long, modified labial palps. Ctenidia is not used by the Protobranchs as food-collecting organs. Most bivalves are filter-feeders (e.g., subclasses Pteriomophia and Heterodonta). Howoever, filter-feeders have well-developed ctenidia. Besides this, the highly specialized carnivore bivalves in the order Septibranchia have their ctenidia modified as septa that help pump water into the body, thereby sucking small crustaceans and other small prey. Like gastropods, bivalves can live in an array of ecological niches such as oysters permanently attach themselves to hard substrates, mussels and ark shells live temporarily attached by bundles of protein fibres called byssus, most of the clams are showing burrowing habitat either in sand or mud. Diverse life style has been marked by different representation from the families. Some of them can bore themselves for life into rock, wood, or other hard substrates. Here is the diagrammatic representation of bivalve showing key identifying features (Figure-13)

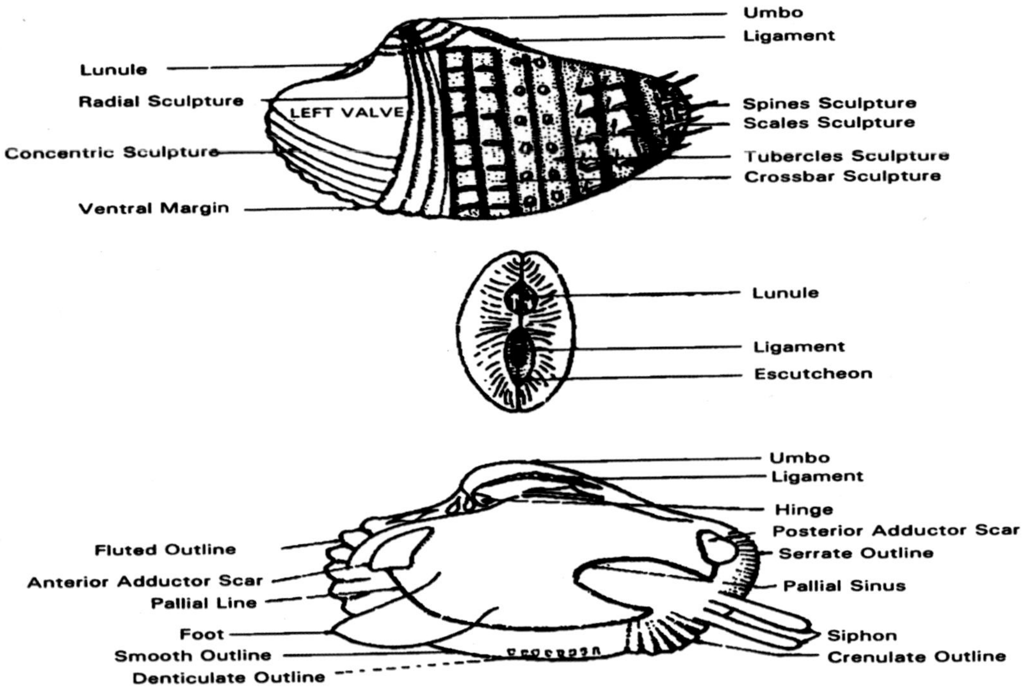

Figure 13: *Shows the available bivalve families in the Chilika lake whereas figure no. 14 showed the outline images common molluscan family of Chilika lake. The term and description related the identification of bivalves is given hereunder.*

Albino - shell lacking normal pigmentation.

Anterior- region situated near the head. In bivalves: region opposite to the siphons, consequently, opposite to the shell sinus.

Beak - same as umbo.

Bivalve - molluscs that have, among other features, shell comprised of 2 halves, or valves.

Byssal - position relative to byssus.

Byssal gap - gap or opening sometimes present on the ventral margin of bivalve shells for passage of byssus.

Byssus - bundle of fibers secreted by some bivalves attaching the animal to the bottom. Mussels, some arks, and pen shells are attached to the substrate by byssus.

Cancellate - cross-barred sculpture. In bivalves, radial and concentric elements will cross to produce a cancellate sculpture.

Chondrophore - depression in spoon-like form housing the internal ligament of some bivalve shells.

Chomata - marginal crenulations in Ostreidae and Gryphaeidae, occurring all around the inner side of valves or only near the hinge, composed of small tubercles or ridgelets on the right valve, and corresponding pits on the left valve.

Compressed - outline of bivalves which are flattened laterally.

Concentric - feature of sculptural elements curving about the umbo in bivalves.

Cord - element of gastropod shell sculpture, usually spirally oriented, thicker than lines.

Cordlet - same as cord.

Corrugated - appearance of surfaces forming wrinkles.

Crenulated - appearance of surfaces which are delicately notched or corrugated. Usually applied to wrinkled shell margin or edge.

Crenulations - notches, or wrinkles which are small and delicate.

Depressed - outline of low, pressed-down gastropod shells. OBS: Term usually applied to some top shells or baby ear.

Dorsal - in bivalves, the region of the hinge.

Elongate - shell with length significantly larger than width.

Equivalve - characteristic of bivalves that have the 2 valves or halves of same size.

Escutcheon - an area of the posterodorsal shell surface near a ligament that is differentiated by sculpture and frequently demarcated by a marginal ridge or furrow.

Excavated - appearance of a hollow, concave surface.

Foliated - characteristic of being leaf-like.

Foot - in bivalves, hatchet-like, expandable structure involved in burrowing.

Glassy - surface resembling glass, vitreous, transparent.

Granulated - surface covered with minute grains, pustules, or beads.

Growth lines - lines on shell surface indicative of alternating periods of growth and rest; sometimes corresponding to seasonal changes.

Hinge - region of the bivalve shell where the two valves are joined together, usually including interlocking teeth and the ligament.

Hinge teeth - projections that interlock on the inner side of the bivalve shell hinge helping to prevent the two

valves from sliding sideways past each other.

Horny - substance that is hardened, proteinaceous; partially or completely forming the ligament, shell periostracum, and possibly other structures.

Incised lines - features of shell sculpture represented by cuts or narrow grooves on the shell surface.

Indented - surface bearing an indentation.

Inequivalve - characteristic of having the two valves (halves) of different size.

Inflated - characteristic of being 'fat', rotund, and frequently lightweight.

Interspaces - spaces between sculptural features, e.g., ribs, costae, or cords.

Juvenile - characteristic of being young, immature, not fully grown.

Knob - large nodule, rounded projection.

Knobbed - surface bearing knobs.

Lamella - thin plate or blade-like projection

Lamellation - same as lamellae.

Ligament - structure that is horny, proteinaceous, acting as a spring tending to keep the valves opened in bivalve shells. Usually situated in the region of the hinge, either internally or externally.

Line - sculptural feature narrowly incised on shell surface.

Lunule - impression on the external side of the hinge, anterior to the umbo, usually heart-shaped.

Mantle - fleshy sheet surrounding vital organs and composed of 2 lobes, one lining and secreting each valve.

Margin - edge of shell.

Nacreous - characteristic of being iridiscent, like mother-of-pearl.

Nodules - projections which are rounded as tubercules.

Nodulose - surface bearing nodules.

Notch - cut or depression, as on a shell margin.

Opalescent - characteristic of being whitish, but with nacreous luster.

Ovate - characteristic of having the form of an egg.

Oval - same as ovate.

Pallial line - fine scar-like impression present internally;in bivalve shells produced by the edge of the mantle.

Periostracum - layer of the outside part of the shell. It is horny and sometimes hair-like.

Plication - same as fold.

Posterior - in bivalves, the region of the shell sinus away from the foot.

Prodissoconch - shell in larval state remaining on the umbonal region of well-preserved bivalve shells.

Radial - structures that are directed away from the umbo toward the shell margin in bivalve shells.

Radiating - same as radial.

Reticulate - feature of shell sculpture consisting of criss-crossed, net-like texture formed by the intersection of lines at right angles.

Reticulated - same as reticulate.

Ribs - strucutural elements forming a well-defined, narrow ridge in gastropod shells. Term usually applied to those elements forming a plane with (or slightly oblique to) shell axis.

Riblets - diminutive of ribs.

Scales - sculptural elements that are small, raised, and plate-like.

Septum - partition found in the internal side of gastropod shells; characteristic of slipper-shells.

Serrated - outline resembling tiny saw teeth.

Shell sinus - embayment on the pallial line of bivalve shells that correspond to the position of the siphons.

Siphon - prolongation of the mollusc mantle used to convey water into or out of the mantle cavity.

Spiral - direction following the coiling of the gastropod shell. Usually applied as a modifier to sculptural terms such as 'spiral cords'.

Striation - fine, repeated lines or furrows on shell surface.

Suture - line or region of junction between two adjacent whorls in the gastropod shell.

Synonym - a scientific name applied to a species that has received an earlier name. Usually, the earlier name is the valid one.

Thread - same as line.

Trigonal - same as triangular.

Umbo (pl. umbones) - projected portion of the hinge. OBS: First-formed part of the bivalve shell.

Varix (pl. varices) - axial sculptural element that is more prominent than a costa, and usually more widely

Spaced; evidence of a growth halt during which a thickened lip develops.

Valve - one half of the bivalve shell.

Ventral - region of the animal opposite the dorsal region; usually region of the foot in bivalves (Source: http://www.fao.org/tempref/docrep/fao/009/y4160e/y4160e04.pdf)

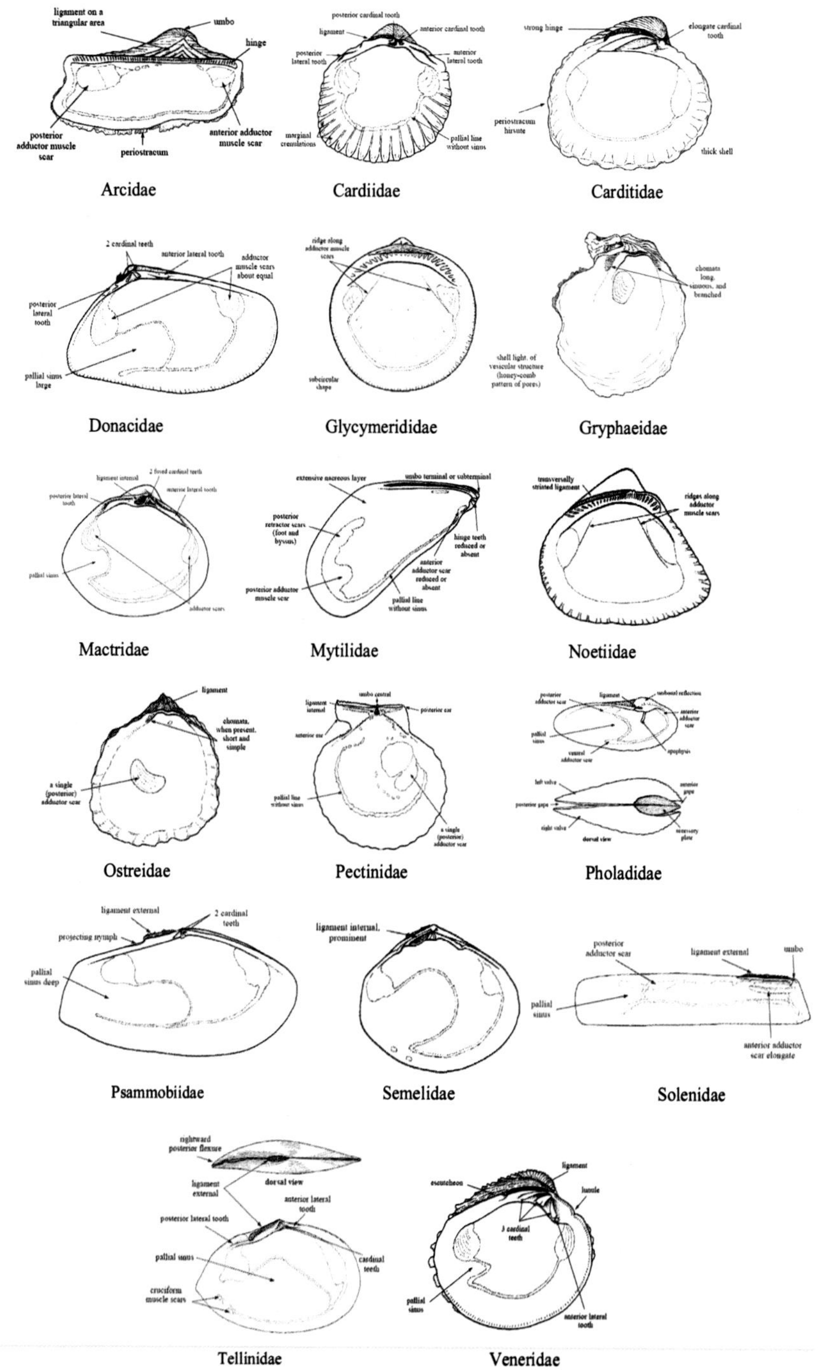

Figure 14: *Outline images common bivalve's family of Chilika.*

Identification of Gastropoda :

It will be entirely not wrong to say that the class gastropoda contains the most species diverse group among the Molluscan kingdom and ranked second after the class insecta of the among the entire invertebrate kingdom. Even this class gastropoda comprises more than half of the total name described for the mollusc in total. However, it is believed that the gastropods are reminiscent of the Cambrian period. The gastropod body is divided two important regions namely (i) head-foot and (ii) mantle (including shell). Mantle contains mantle cavity, and visceral mass. Generally, the muscular foot of the gastropod is used for locomotion, to crawl, to attach, and for burrowing exercise. Whereas the head part includes sensory organs such as tentacles and eyes. Few gastropods also composed with nerve ganglia and connectives. The mantle is formed because of the secretion of the calcium carbonate externally from the mantle cavity. Mantle cavity is that structure which lies between the head-foot and the mantle proper. It further contains the ctenidium (or ctenidia), osphradium, anus, nephridiopore, and external genitalia. The posterior direction of the shell contains the visceral mass. This also contains gonads, digestive gland, heart, kidney, and part of the alimentary system.

Exceptionally diversified food habits are marked among the gastropods. They are the predators, scavengers, filter- and deposit-feeders, macro- and micro-herbivores. Furthermore drilling technique may be adopted by some of the gastropod in order to reach for their source of feeding by using the radula. From the standpoint of reproduction, gastropods may be dioeciously or hermaphroditic (simultaneous or protandric), and may perform internal or external fertilization. Gastropods undergoes through pelagic larval development for different time duration i.e. it may be ranged from a couple of hours to a few months. But the most noteworthy information is that some species may surpass the pelagic larval developmental stage and may be followed the intracapsular (direct) development as an alternative method of growth. Since, gastropods are also extremely diversified; they can inhabit both terrestrial and aquatic environments. In the marine mileu their habitat is ranged in a diverse condition. It can be ranged from the upper littoral zone to the abyssal depths.

The figure 15 showing the different parts of the gastropod. The family wise identification remark of some gastropod shells are given in the figure16. The terms and the conditions related to the identification of the gastropoda is given hereunder.

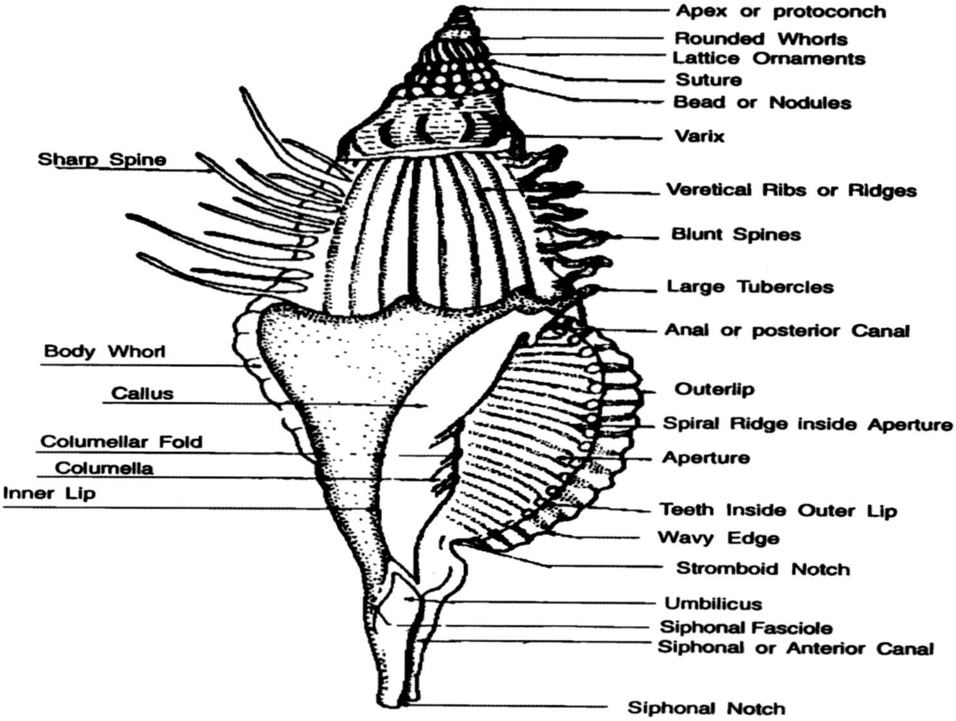

Figure 15: *Hypothetical shell structures of a gastropod showing different identifying marks.*

Albino - shell lacking normal pigmentation.

Anterior - region situated near the head. In gastropods: in front.

Anterior canal - expansion looking like a groove or a tube and serving to protect the siphon in gastropod shells.

Aperture - opening in gastropod shells.

Apertural - position relative to the aperture of gastropod shells.

Apex - extremity of a gastropod shell opposite to the anterior region; part of the shell built in earlier life.

Apical - situated at or near the apex of a gastropod shell.

Axial - direction forming a plane with main shell axis in gastropods.

Basal - position relative to shell base.

Base - part of the gastropod situated in opposition to the apex.

Bottom of the shell - same as base.

Body whorl - most anterior whorl of the gastropod shell, last and largest whorl.

Callus - thickening of the shell, secondary, smooth, sometimes glazed, usually secreted on the parietal region of the columella.

Cancellate - feature of cross-barred sculpture of some gastropod shells consisting of axial and spiral elementsbof same intensity crossing at right angles.

Columella - column or pillar located on the centre of a gastropod shell.

Cord - element of gastropod shell sculpture, usually spirally oriented, thicker than line.

Cordlet - narrow cord, thicker than line.

Corrugated - appearance of surfaces forming wrinkles.

Crenulated - appearance of surfaces that are delicately notched or corrugated. Term usually applied to wrinkled shell margin or edge.

Crenulations - notches, or wrinkles that are small and delicate.

Denticles - features of sculpture elements looking like small teeth-like projections.Term usually applied to features seen on the internal part of the aperture.

Depressed - outline of low, pressed-down gastropod shells. Term usually applied to some top shells.

Dorsal - region opposite to the foot in gastropods.

Egg-ribbon - same as ribbon.

Elongate - shell with length significantly larger than width.

Excavated - appearance of a hollow, concave surface.

Fold - ridge spiralling on columella.

Foliated - characteristic of being leaf-like.

Foot - in gastropods, fleshy, sole-like, muscular part of body involved in locomotion.

Furrow - groove in longitudinal direction found on the dorsal region of, among other shells, cowries and Triviidae.

Fusiform - characteristic of being spindle-shaped.

Glassy - surface resembling glass, vitreous, transparent.

Globular - shape resembling a sphere or a ball.

Globose - same as globular.

Granulated - surface covered with minute grains, pustules, or beads.

Growth lines - lines on shell surface indicative of alternating periods of growth and rest; sometimes corresponding to seasonal changes.

Horny - substance that is hardened and proteinaceous; present in or completely forming the gastropod operculum and shell periostracum.

Incised lines - features of shell sculpture represented by cuts or narrow grooves on the shell surface.

Indentation - cut or notch on shell edge or parietal region.

Indented - surface bearing an indentation.

Interspaces - spaces between sculptural features, such as ribs, costae, or cords.

Juvenile - characteristic of being young, immature, not fully grown.

Keyhole - apical orifice in some limpets.

Knob - large nodule, rounded projection.

Knobbed - surface bearing knobs.

Lamella (pl. lamellae) - thin plate or blade-like projection.

Lamellation - same as lamella.

Ligament - structure that is horny and proteinaceous, acting as a spring tending to keep the valves opened in bivalve shells. Usually situated in the region of the hinge, either internally or externally.

Line - sculptural feature narrowly incised on shell surface.

Lip - edges of the outer surface of the aperture in the gastropod shell.

Longitudinal - direction parallel to the largest dimension of the shell or mollusc.

Nacreous - characteristic of being iridiscent, like mother-of-pearl.

Nodules - projections that are rounded as tubercules.

Nodulose - surface bearing nodules.

Notch - cut or depression on any margin, canal, or on the gastropod aperture.

Opalescent - characteristic of being whitish, but with nacreous luster.

Operculum - trapdoor or plate which closes the aperture of gastropod shells and isolates the snail from its surrounding environment. Opercula can be horny ('soft', brownish) or calcareous ('hard', usually whitish).

Outer lip - edge of the external part of the aperture away from the shell axis.

Ovate - characteristic of having the form of an egg.

Oval - same as ovate.

Parietal - region of the internal part of the aperture, usually set apart by differences in surface texture and/or coloration.

Parietal shield - parietal region when markedly different from the remainder of the adjacent shell area.

Periphery - region of the outermost part of any given whorl on the gastropod shell. The shell periphery is therefore the greatest circumference of the gastropod shell.

Periostracum - layer of the outside part of the shell. It is horny and sometimes hair-like.

Peristome - aperture rim or periphery.

Plication - same as fold.

Posterior - region away from the siphonal canal, near the apex, in gastropods; in bivalves, the region of the shell sinus, away from the foot.

Posterior canal - canal of small size or notch opposite to the siphonal canal on the aperture of the gastropod shell.

Protoconch - larval shell remaining on the apex of well-preserved gastropod shells.

Radial - structures that are directed away from the apex toward the shell margin, in limpets.

Radiating - same as radial.

Reticulate - feature of shell sculpture consisting of criss-crossed, net-like texture formed by the intersection of lines at right angles.

Reticulated - same as reticulate.

Ribbon - surface consisting of an aligned sequence of egg-cases.

Ribs - structural elements forming a well-defined, narrow ridge in gastropod shells. Term usually applied to those elements forming a plane with (or slightly oblique to) shell axis.

Riblets - diminutive of ribs.

Scales - sculptural elements that are small, raised, and plate-like.

Septum - partition found in the internal side of gastropod shells; characteristic of slipper-shells.

Serrated - resembling tiny saw teeth.

Shoulder - angled region of the whorls of gastropod shells.

Siphon - prolongation of the gastropod mantle used to convey water into the mantle cavity.

Siphonal canal - projection of the anterior region shell in tubular form protecting the anterior siphon.

Snails - common name of gastropods.

Spiral - direction following the coiling of the gastropod shell. Term usually applied as a modifier to sculptural terms such as 'spiral cords'.

Spire - series of successive whorls in a gastropod shell, with exception of the last one.

Spire angle - angle formed by the lines defined by the outermost points on both sides of the spire.

Striation - fine, repeated lines or furrows on shell surface.

Suture - line or region of junction between two adjacent whorls in the gastropod shell.

Synonym - a scientific name applied to a species that has received an earlier name. OBS: Usually, the earlier name is the valid one.

Thread - same as line.

Top of the shell - same as apex.

Turbinate - form that looks top-shaped, tapering evenly from base to apex.

Turreted - form that looks tower-shaped, elongate.

Umbilicus - cavity at base of gastropod shells.

Uncoiled - gastropod shell that lacks coiling.

Varix - axial sculptural element that is more prominent than a costa, and usually more widely spaced; evidence of a growth halt during which a thickened lip develops (plural: varices).

Ventral - region of the animal opposed to the dorsal region; region of the foot in gastropods.

Whorl - a complete turn or coil of the gastropod shell (Source:www.fao.org/tempref/docrep/fao/009/y4160e/y4160e08.pdf).

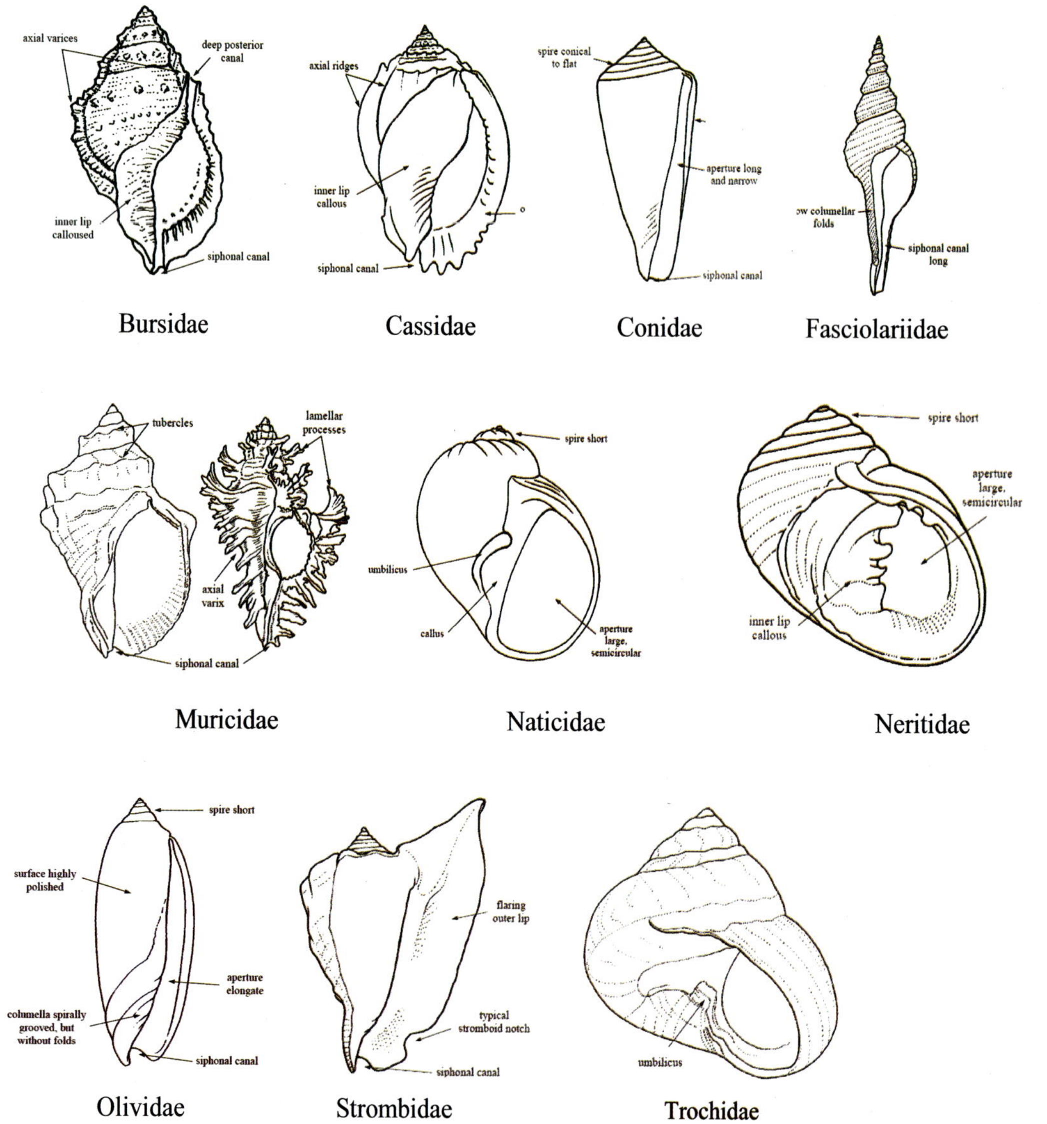

Figure 16: *Outline images common Gastropod family of Chilika lake.*

4

Checklist of Molluscan fauna of Chilika

Molluscan Species diversity:

From table no. 5 the detailed list of observed molluscan families were described from Chilika lagoon those were collected during the study period. April 2007 to March 2010. A total of 147 molluscan species were identified belonging to 3 classes 20 order, 67 families. Highest species were recorded among the class gastropod (82 sps.) belonging to 9 orders and 40 families followed by class bivalves (64 sps.) under 10 order and 26 families. Least species contribution was seen by the polyplacophorans representing 1sps. namely *Acanthochitona mahensis* comprising to 1 order and 1 family only. The details of the molluscan family recorded during this study are hereunder:

Table 5: List of Molluscan Families recorded during 2007 to 2010

Sl. no.	Gastropod	Bivalves	Polyplacophora
1	Family Acteocinidae	Family Anomiidae	Family Chitonidae
2	Family Ampullariidae	Family Arcidae	
3	Family Aplysiidae	Family Cardiidae	
4	Family Arminidae	Family Carditidae	
5	Family Babyloniidae	Family Cultellidae	
6	Family Bithyniidae	Family Cuspidariidae	
7	Family Bursidae	Family Cyrenidae	
8	Family Cancellaridae	Family Donacidae	
9	Family Cassidae	Family Glauconomidae	
10	Family Conidae	Family Glycymerididae	
11	Family Epitoniidae	Family Gryphaeidae	
12	Family Fasciolariidae	Family Laternulidae	

13	Family Ficidae	Family Mactridae	
14	Family Haminoeidae	Family Mytilidae	
15	Family Harpidae	Family Noetiidae	
16	Family Hydrobiidae	Family Ostreidae	
17	Family Litiopidae	Family Pholadidae	
18	Family Littorinidae	Family Placunidae	
19	Family Lymnaeidae	Family Psammobiidae	
20	Family Melongenidae	Family Semelidae	
21	Family Muricidae	Family Solenidae	
22	Family Nassariidae	Family Tellinidae	
23	Family Naticidae	Family Trapeziidae	
24	Family Neritidae	Family Ungulinidae	
25	Family Olividae	Family Unionidae	
26	Family Planorbidae	Family: Veneridae	
27	Family Potamididae		
28	Family Pyramidellidae		
29	Family Rostellariidae		
30	Family Scaliolidae		
31	Family Solariellidae		
32	Family Stenothyridae		
33	Family Strombidae		
34	Family Thiaridae		
35	Family Tonnidae		
36	Family Trochidae		
37	Family Turridae		
38	Family Turritellidae		
39	Family Viviparidae		
40	Family Xenophoridae		

Bivalves

Among bivalves, out of nine the order Veneroida is comprised of 41 species under 13 families & becomes the most dominant comprised of 66 % of the total bivalve species observed in Chilika during this study (Table-6) . The families those are fall under the order Veneroida are Veneridae, Cardiidae, Carditidae, Cultellidae, Cyrenidae, Donacidae, Glauconomidae, Mactridae, Psammobiidae, Semelidae, Tellinidae, Trapeziidae, Ungulinidae. Above all, the family Veneridae was comprised of 12 species alone contributed a major percentage to the total bulk of macrobenthic fauna.

A number of species of the class bivalves are found dominant in the lagoon bed of Chilika. These are mostly found from the outer channel area, southern sector, Central sector and Palur Channel connected point. Some of the dominant species are those of *Meretrix meretrix* Linnaeus, 1758, *Meretrix casta* Gmelin, 1791, *Sunetta scripta* Linnaeus, 1758, *Theora opalina* Preston 1914, *Macoma birmanica* Philippi, *Tellina texturata* Sowerby in Reeve's 1867, *Silliqua aculata* Linnaeus, 1756, *Mactra mera* Reeve, 1854, *Tegillarca granosa* Linnaeus, 1758, *Brachidontes striatulus* Hanley, 1843, *Brachidontes undulates* Dunker, 1857, *Perna viridis* Linnaeus, 1758, *Solen kempi* Preston, *Crassostrea cuttackensis* Newton & Smith, 1912, *Saccostrea cucullata* Born 1778, *Placuna placenta* Linnaeus, 1758, *Hyotissa hyotis* Linnaeus, 1758, *Cuspidaria chilkaensis* Preston . Out of these *Crassostrea gryphoides Crassostrea cuttackensis, Saccostrea cucullata, Placuna placenta, Hyotissa hyotis, Sunetta scripta, Meretrix meretrix, Perna viridis* are exclusively found in the high salaine region where the salinity is above22-35 ppt. Species like *Theora opalina, Tegillarca granosa, Brachidontes undulates, Brachidontes striatulus, Macoma birmanica* are exclusively of brackish water in nature. One fresh water species namely *Lamellidens marginalis* have the ability to produce pearl, commonly seen in the fresh water dominated region of northern sector and central sector where as one wood boring bivalve *Martesia striata* is frequently occurring in the wooden boats, different shoots of mangroves and other wooden structures are found in the lake premises. The windowpane oyster *Placuna placenta* was recorded from the sandy bottom of outer channel area where as species namely *Brachidontes striatulus, Perna viridis, Hyotissa hyotis, Crassostrea cuttackensis, Saccostrea cucullata,* are acting as potential biofoulers.

Gastropoda:

Gastropods dominate the lagoon bed. Most of them are comprised of the order Neogastropoda which includes 30 species belonging to 10 families measuring a share of 37 % of the total species of gastropod. The second most prevailing group was observed in the form of Caenogastropod which contributes 23 % of the gastropod having 19 species grouped under 10 families.

Both of these orders jointly contributed 49 species (30 from Neogastropod + 19 from Caenogastropod). Overall contribution when calculated it brings out a voluminous share of 60 % of the total.

The order Neogastropoda was predominant in the study area (Table 6). Ten families namely Cancellaridae, Conidae, Fasciolariidae, Harpidae, Melongenidae, Muricidae, Nassariidae, Olividae, Pyramidellidae, Turridae are found under this order. Whereas eight families belonging to order Caenogastropoda such as Epitoniidae, Ficidae, Litiopidae, Potamididae, Scaliolidae, Thiaridae, Turritellidae, Babyloniidae were recorded as second predominant group in the lake substratum of Chilika.

The observed dominant species of gastropods found in the study period, are those of *Babylonia spirata* Linnaeus 1758, *Bellamya bengalensis* Lamarck 1822, *Bulia vittata* Linnaeus 1767, *Cerithideopsilla cingulata* Gmelin1791,*Cerithideopsis costata* da Costa, 1778, *Indothias lacera* Born 1778,*Littorina littorea* Linnaeus 1758, *Murex tribulus* Linnaeus 1758, *Natica tigrina* Röding 1798, *Nassarius stolatus* Gmelin 1791, *Neritina smithii* Wood 1828, *Oliva oliva* Linnaeus 1758, *Olivancillaria gibbosa* Born 1778, *Phalium areola* Linnaeus 1758, *Pila globosa* Swainson 1822, *Stenothyra blanfordiana* Nevill 1880, *Tarebia acula* Muller 1774, *Telescopium telescopium* Linnaeus, 1758, *Turritella aculate* Reeve 1849, *Umbonium vestiarium* Linnaeus 1758.

Table 6: Systematic account showing the observed molluscan species of Chilika lake

Phylum: Mollusca	
Class: BIVALVIA	
Order: Veneroida	
Family: Veneridae	
Sl.no	**Genus/species**
1	*Meretrix meretrix* Linnaeus, 1758
2	*Meretrix casta* Gmelin, 1791
3	*Sunetta scripta* Linnaeus, 1758
4	*Sunetta meroe* Linnaeus, 1758
5	*Clementia papyracea* Gmelin, 1791
6	*Tivela dilwyni* Deshayes, 1853
7	*Dosinia fibula* Reeve,1850
8	*Venus pinguis* Chemnitz, 1782
9	*Paphia aculate* Born1778

10	*Timoclea aculate* Sowerby II, 1853
11	*Timoclea acula* Hanley, 1845
12	*Pitar erycina* Linnaeus, 1758
	Family Cyrenidae
13	*Corbicula striatella* Deshayes,1854
	Family Glauconomidae
14	*Glauconome sculpta* Subba Rao 1977
	Family Tellinidae
15	*Tellina tenuis* da Costa, 1778
16	*T. texturata* Sowerby in Reeve's 1867
17	*T.barhampurensis* Preston
18	*Macoma birmanica*
	Family Semelidae
19	*Theora opalina* Preston 1914
20	*Cumingia hinduorum* Annandale and Kemp, 1916
	Family Donacidae
21	*Donax aculate* Gmelin 1791
22	*Donax scortum* Linnaeus
	Family Trapeziidae
23	*Trapezium sublaevigatum* Lamarck
	Family Psammobiidae
24	*Gari mahosaensis* Preston
25	*Sanguinolaroia accuminata*
	Family Cultellidae
26	*Cultellus subellipticus* Dumker
27	*Neosolen aquaedulcioris* Ghosh
28	*Silliqua aculat* Linnaeus, 1756
	Family Mactridae
29	*Mactra stultorum* Linnaeus, 1758
30	*M.grandis* Gmelin
31	*M.luzonica* Deshayes
32	*M.mera* Deshayes
33	*M.turgida* Gmelin
34	*Spisula annandalei* Preston

	Family Ungulinidae
35	*Diplodonta barhampurensis* Preston
36	*Diplodonta sataparensis* Preston
37	*Felania annadalei* Preston
38	*Felania ovalis* Preston
	Family Carditidae
39	*Cardites aculata* Linnaeus 1758
	Family Cardiidae
40	*Cardium coronatum* Schroeter, 1786
41	*Cardium latum* Born, 1778
	Order Arcoida
	Family Noetiidae
42	*Striarca lacteal* Linnaeus, 1758
	Family Arcidae
43	*Tegillarca granosa* Linnaeus, 1758
44	*Tegillarca rhombea* Born, 1778
45	*Scapharca deyrollei* Jousseaume, 1893
	Family Glycymerididae
46	*Glycymeris tenuicostata* Reeve, 1843
	Order Mytiloida
	Family Mytilidae
47	*Brachidontes striatulus* Hanley, 1843
48	*Brachidontes undulates* Dunker, 1857
49	*Modiolus modiolus* Linnaeus, 1758
50	*Brachidontes modiolus* Linnaeus, 1767
51	*Perna viridis Linnaeus*, 1758
52	*Perna perna* Linnaeus, 1758
	Order Euheterodonta
	Family Solenidae
53	*Solen annandalei* Preston, 1915
54	*Solen kempi* Preston
	Order Ostreoida
	Family Ostreidae
55	*Crassostrea cuttackensis* (Newton & Smith, 1912)

56	*Crassostrea gryphoides*
57	*Saccostrea cucullata* Born 1778
	Order Pectinoida
	Family Anomiidae
58	*Anomia achaeus* Gray 1839
	Family Placunidae
59	*Placuna placenta* Linnaeus, 1758
	Family Gryphaeidae
60	*Hyotissa hyotis* Linnaeus, 1758
	Order Anomalodesmata
	Family Cuspidariidae
61	*Cuspidaria chilkaensis* Preston
	Order Myoida
	Family Laternulidae
62	*Laternula navicula* Reeve, 1863
	Order Unionoida
	Family Unionidae
63	*Lamellidens marginalis* Lamarck,
	Order Myida
	Family Pholadidae
64	*Martesia striata* Linnaeus, 1758
Class: GASTROPODA	
Order: Vetigastropoda	
Family: Trochidae	
1	*Umbonium vestiarium* Linnaeus, 1758
2	Turbo brunneus
	Family Solariellidae
3	Solariella obscura Couthouy,1838
	Order Cycloneritimorpha
	Family Neritidae
4	*Nerita balteata* Reeve, 1855
5	*Neritina smithii* Wood, 1828
	Order Architaenioglossa
	Family Viviparidae

6	*Bellamya bengalensis* Lamarck, 1822
	Family Ampullariidae
7	*Pila globosa* Swainson, 1822
	Order Caenogastropoda
	Family Turritellidae
8	*Turritella aculate* Reeve, 1849
9	*Turritella columnaries* Kiener, 1843
10	*Turritella aculate* Reeve, 1849,
11	*Turritella terebra* Linnaeus, 1758
	Family Thiaridae
12	*Tarebia granifera* Lamarck, 1822
13	*Tarebia acula* Muller 1774
14	*Tarebia aculat* Gray
	Family Potamididae
15	*Cerithideopsilla cingulata* Gmelin 1791
16	*Cerithideopsis costata* da Costa, 1778
17	*Telescopium telescopium* Linnaeus, 1758)
	Family Epitoniidae
18	*Epitonium hamatulae* Preston, 1915
	Family Litiopidae
19	*Alaba blanfordi* Adams, 1862
	Family Scaliolidae
20	*Finella virgata* Philippi, 1849
	FamilyBabyloniidae
21	*Babylonia spirata* Linnaeus, 1758
22	*Babylonia zeylanica*
	Family Ficidae
23	*Ficus aculata* Roeding, 1798
24	*Ficus gracilis* Sowerby, 1825
	Family Rostellariidae
25	*Tibia delicatula*
	Family Xenophoridae
26	*Stellaria solaris* Linnaeus, 1764
	Order Littorinimorpha

	Family Naticidae
27	*Natica gualteriana* Récluz, 1844
28	*N. vitellus* Linnaeus, 1758
29	*N. tigrina* Röding, 1798
30	*Eunaticina coarctata* Reeve
	Family Littorinidae
31	*Littorina littorea* Linnaeus, 1758
32	*L. melanostoma* Gray, 1839
33	*L. aculate* Gray, 1839
	Family Cassidae
34	*Phalium areola* Linnaeus, 1758
	Family Bithyniidae
35	*Bithynia sistanica* Annandale & Prashad, 1919
	Family Stenothyridae
36	*Stenothyra blanfordiana* Nevill, 1880
37	*S. minima* Sowerby I, 1837
38	*Gangetia miliacea* Nevill, 1880
	Family : Tonnidae
39	*Tonna dolium* Linnaeus, 1758
40	*Tonna sulcosa*
41	*Tonna allium*
	Family: Strombidae
42	*Strombus* Linnaeus, 1758
	Order Neogastropoda
	Family Muricidae
43	*Indothias lacera* Born, 1778
44	*Thias bufo*
45	*Cronia contracta* Reeve, 1846
46	*Murex tribulus* Linnaeus, 1758
47	*Chicoreus virginus*
48	*Chicoreus microphyllus* Lamarck, 1816
	Family Melongenidae
49	*Pugilina cochlidium Linnaeus, 1758*
	Family Nassariidae

50	*Hyanassa aculata* Say 1822
51	*Nassarius stolatus* Gmelin, 1791
52	*N. foveolatus* Dunker, 1847
53	*N. subconstrictus* Sowerby, 1899
54	*N. jacksonianus Quoy & Gaimard, 1833*
55	*N. orissaensis* Preston, 1914
56	*Bulia vittata* Linnaeus, 1767
	Family Olividae
57	*Oliva oliva* Linnaeus, 1758
58	*Olivancillaria gibbosa* Born, 1778
	Family Cancellaridae
59	*Cancellaria elegans* GB Sowerby I, 1822
	Family Pyramidellidae
60	*Odostomia chilkaensis* Preston 1914
61	*Pyrgulina ecclesia* Annandale & Kemp, 1916
62	*Pyrgulina humilis* Preston, 1905
63	*Pyrgulina nadiensis* Preston, 1905
64	*Turbonilla rambhaensis* Preston, 1905
65	*Turbonilla aculat* Linnaeus, 1758
66	*Linopyrga aculat* Saurin, 1959
67	*Quirella humilis* Preston, 1905
	Family Conidae
68	*Conus hyaena* Hwass in Bruguière, 1792
69	*Conus inscriptus* Reeve, 1843
	Family Fasciolariidae
70	*Fusinus colus*
	Family Harpidae
71	*Harpa davidis* Roding, 1798
	Family Turridae
72	*Turricula javana*
	Order Cephalaspidea
	Family Haminoeidae
73	*Haminoea crocata* Pease, 1860
	Family Acteocinidae

74	*Tornatina estriata* Preston 1914
	Order Hygrophila
	Family Lymnaeidae
75	*Radix luteola* Lamarck, 1822
76	*Lymnaea catascopium* Say, 1817
	Family Planorbidae
77	*Indoplanorbis exustus* Deshayes, 1834
	Family Arminidae
78	*Armina maculata* Rafinesque, 1814
	Order Anaspidea
	Family Aplysiidae
79	*Aplysia dactylomela* Rang, 1828
	Family Hydrobiidae
80	*Peringia ulvae* Pennant, 1777
	Family Bursidae
81	*Bufonaria echinata* Link, 1807
82	*Bursa spinosa* Lamarck, 1843
Class : POLYPLACOPHORA	
Order: Chitonida	
Family: Chitonidae	
1	*Acanthochitona mahensis* Winckworth, 1927

5

Bivalves: Pictorial Guide

During the present study a total of 64 species of bivalves were recorded. Most of the species are common and recorded previously where as few of the newly occurring species were also recorded. Most of the new records were made from the outer channel area. The taxonomic classification, description, distribution details of certain bivalves are given here under.

Anomia achaeus

Phylum : Mollusca

Class: Bivalvia

Order : Pectinoida

Family : Anomiidae

Genus and species : *Anomia achaeus* Gray 1839

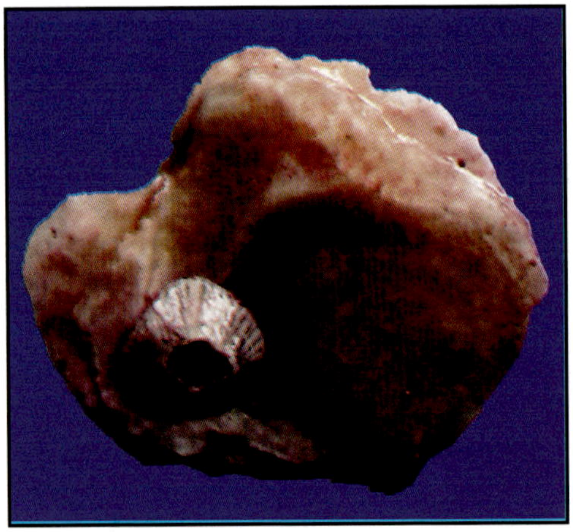

Anomia achaeus

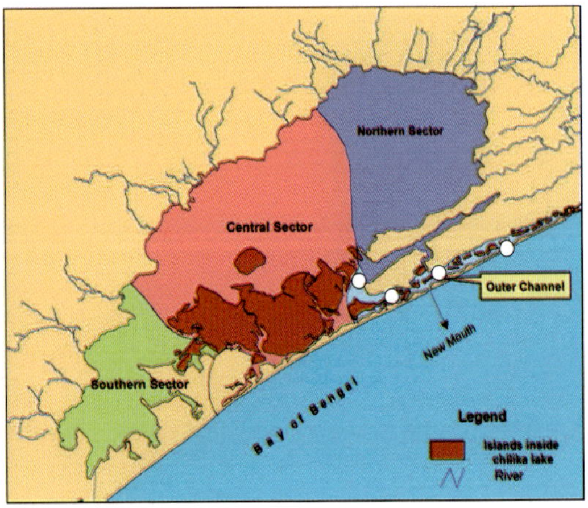

Distribution in Chilika lagoon

Description Size: 20 mm; length 16.40-25.10 mm; width 21.50-28.15 mm. Irregular shell, thin, fragile, translucent, usually attached to substratum. Valve unequal, right valve flat and attached to the bedrock, left valve arched, infalted, sculptured surface.

Body Color: Golden yellow.

Distribution in Chilika lake : Old mouth (Arakhkuda), New mouth (Shipakuda), Satapada, Rambharatia Is.

Distribution in India : Gujarat, Gulf of Kachchh, Andhra Pradesh, Godavari Estuary, West Bengal, Digha Coast, Orissa, Chilika Lagoon, West Bengal: Digha; Medinapur; Maharashtra, Gujarat, Maharashtra: Mumbai Kerala.

Abroad : Karachi, Persian Gulf , Gulf of Aden

Remark: It was found in associated with other molluscan representatives such as *Crassostrea* sp., *Saccostrea cucullata* and gastropod *Telescopium telescopium*. It is also found on the rocks, crevices, cemented structure and other hard structures often visible in the low tide periods. Previously it's distribution is restricted to outer channel area only but now it has spread to those areas where minimum salinity id 20 ppt or more. It is a potent bio-fouler, also recorded from the jetty pillars and base of the boats, trawlers and barz.

Cardita antiquata

Phylum : Mollusca

Class: Bivalvia

Order :Veneroida

Family : Carditidae

Genus and species : *Cardita antiquata* Linnaeus 1758

Cardita antiquataa

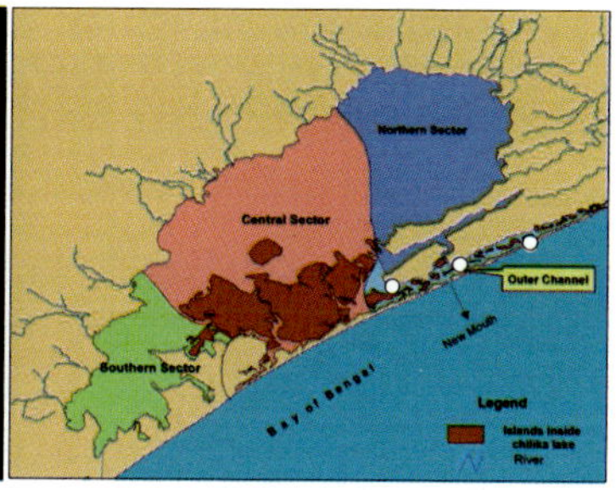

Distribution in Chilika lagoon

Description Size: Length 7.3 - 42.20 mm; height 7.24-35.30 mm. Shell inequilateral, either cordiform or mytiliform. Hinge with laterals rudimentary, not distant; cardinals, large, grooved; margin crenulate

Distribution in Chilika : Outer channel area, Arakhkuda (old inlet), New lagoonal inlet,

Distribution in India: Odisha : Chilika lake, Rushikulya estuary Chatrapur, Gopalpur, Puri, Konark, Chandrabhaga, Paradwip, **Tamil Nadu:** Kundugal Point; Chennai; Pondicherry; Krusadai; Tuticorin; **Maharashtra**; Mumbai, **Gujarat**: Gulf of Kachchh,**Goa**

Abroad: Sri Lanka

Remark: Very less information is available.

Cardium coronatum

Phylum : Mollusca

Class: Bivalvia

Order :Veneroida

Family : Cardiidae

Genus and species : *Cardium coronatum* Schroeter, 1786

Cardium latum Born, 1778

Cardium coronatum

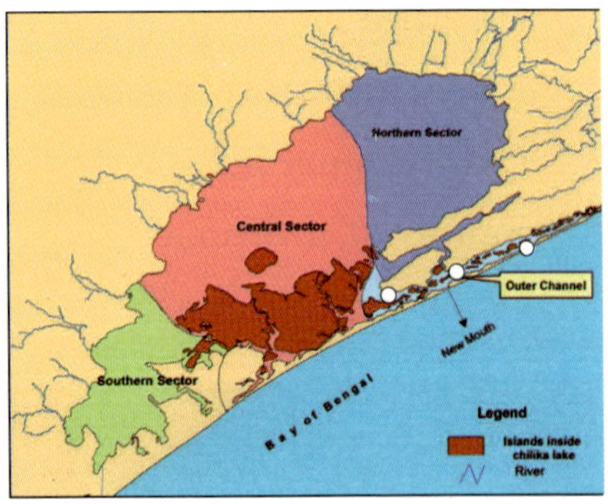

Distribution in Chilika lake

Description Size: 7.57-33.21 mm. Shell thin, height is almost equal to length; rounded in shape; ribs rounded; valves more inflated.

Distribution in Chilika lake : Satapada, Shipakuda, new inlet, old inlet (Arakhkuda), also recorded earlier from Satapad and Manikpatna

Distribution in India: In **Odisha**: Budhabalang River Mouth; Baleshwar, Chandipur, Paradwip, False Point, Konark, Chilika lake, Rushikulya estuary, Gopalpur, **Andhra Pradesh**: Vishakhapatnam, **Tamil Nadu** : Chennai; Krusadai; Palk Bay; Mandapam, Maharashtra: Devgarh; Mumbai, **Gujarat**: Gulf of Kachchh, **Andaman and Nicobar Islands**.

Abroad : Persian Gulf, Sri Lanka

Remark: Extensively used in lime formation . Mostly seen during summer. Abundance had been increased after the opening of the new lagoonal inlet.

Clementia papyracea / *vatheleti*

Phylum : Mollusca

Class: Bivalvia

Order :Veneroida

Family:Veneridae

Genus and species: *Clementia papyracea*

Clementia vatheleti Mabille

Clementia vatheleti

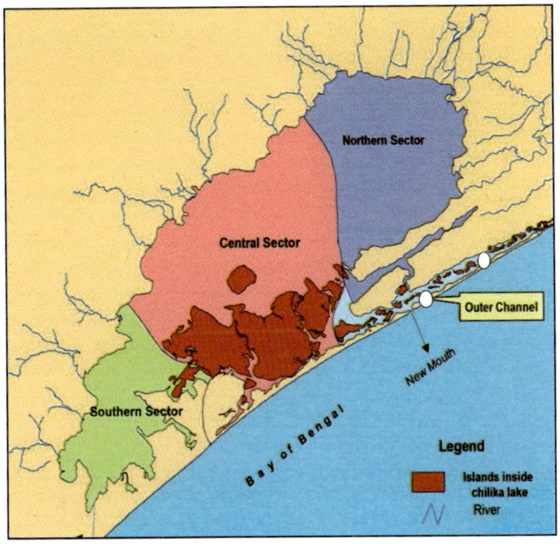

Distribution in Chilika lagoon

Description: Shell thin, fragile; lunule indistinct, depressed; laterals absent; surface with irregular concentric ribs. Size: Length 5.2-12.8 mm; height 5.22-13.53 mm; thickness 2.850-10. Mm

Body Colour: Pale white or yellow in color.

Distribution in Chilika : Presently, dead shells of this specimen is abundantly collected from outer channel area, Earlier this species was recorded from Chadheihaga is. off Balugaon, Nalabana bird Is., Rambha Bay and outer channel area.

Distribution in India: Andhra Pradesh, Kakinada Bay, West Bengal, Port Canning

Remark: Dead shells are observed in large scale throughout the year. Live shell is seen during the winter season.

Corbicula striatella

Phylum : Mollusca

Class: Bivalvia

Order :Veneroida

Family : Cyrenidae

Genus and species: *Corbicula striatella* Deshayes,1854

Corbicula striatella

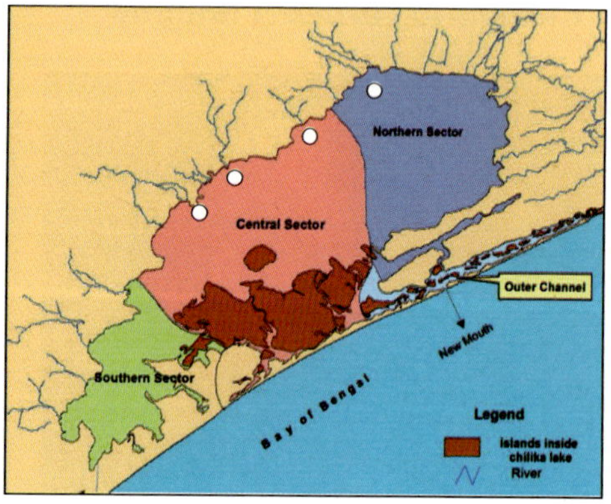

Distribution in Chilika lagoon

Description Size: 8.20-12.22 mm**Distribution in Chilika :** Fresh water dominated regions of northern sector and central sectors, river mouth regions.

Distribution in India: Bihar, Haryana, Jharkand, Madhya Pradesh, Rajasthan, Uttaranchal, Uttar Pradesh, West Bengal

Abroad: Bangladesh; Myanmar; Nepal; Sri Lanka.

Remark*:* It is a common freshwater bivalve species often located in rivers, streams and lakes, prefers to live in fine gravel, sand or mud substratum; mostly confined to lowland rivers. It is used as food in many parts of Nepal, India and

Bangladesh. In Chilika, it embeds in the river mouth bed in the Kalupadaghat region, in Sorana, Tinimuhani, Bhusandapur region etc. It provides a great socio-economic support to the local fishermen since they used it as food particles for prawn culture. Dead shells are used for the lime formation.

Crassostrea cuttackensis

Phylum : Mollusca

Class: Bivalvia

Order : Ostreoida

Family : Ostreidae

Genus and species : *Crassostrea cuttackensis* Newton & Smith, 1912

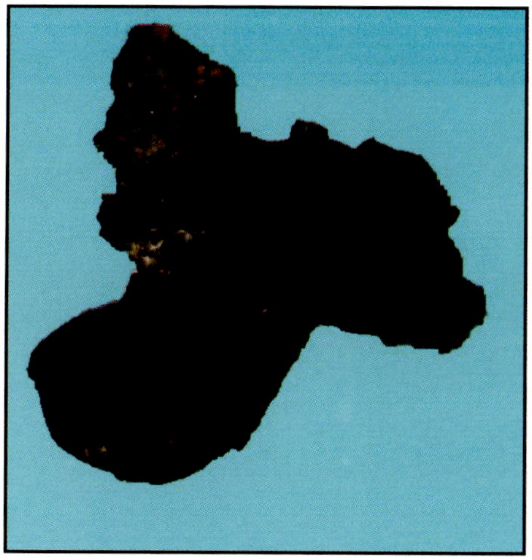

Crassostrea cuttackensis

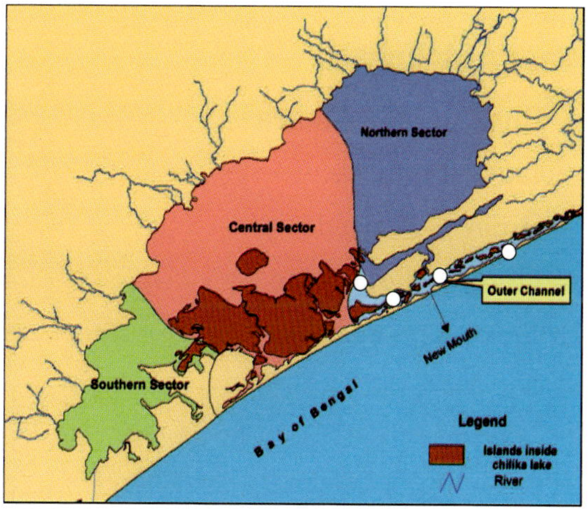

Distribution in Chilika lagoon

Description Size: 19.25-66.62 mm; length 130.40-290 mm; width 56-145 mm; height 24.25-83.75 mm, Shell large exceeding 50 mm in length, heavy, bulky, lower valve concave, margin not crenulate, no tubercles on either side of umbonal groove; variable shape. The shape, size and even the impression of muscle scar change due to the environmental condition.

Body Colour: Dark brown in colour.

Distribution in Chilika lake : Old mouth (Arakhkuda), New mouth (Shipakuda), Satapada, Rambharatia Is.

Distribution in India : Odisha, Mahanadi Estuary; Chilika Lake, Gopalpur; Chhatrapur; Rushikulya estuary; Hukitola; Chandipur; **West**

Bengal: Jharkhali; North 24 Parganas; Sundarbans; Champta. **Andhra Pradesh,** Chittivalasa; **Gujarat** : Gulf of Kachchh, **Karnataka**, **Kerala**, Maharashtra, **Tamil Nadu** : Ennore Backwaters, Gulf of Mannar; Krusadai; Shingle Island, Kerla, Kochi; Kozhikode, Karnataka: Karwar, **Maharashtra**: Mumbai, **Lakshadweep**, **Andaman** and **Nicobar** Islands.

Abroad: Bangladesh , Malaysia.

Remark: This species is estuarine and intertidal can live within the salinity range of 20 ppt to 35 ppt embedded in sandy substratum. In many parts it used as food item for human consumption and also used as bait to catch fish and shell fish. This species is also used for the extraction of lime. They crawl on the substratum in order to feed on settled planktonic materials. This is also used as an indicator of heavy metal pollution, oil pollution in the many parts of the world.

Donax pulchellzls

Phylum : Mollusca

Class: Bivalvia

Order :Veneroida

Family : Donacidae

Genus and species : *Donax pulchellzls* Gmelin 1791*

Donax pulchellzls

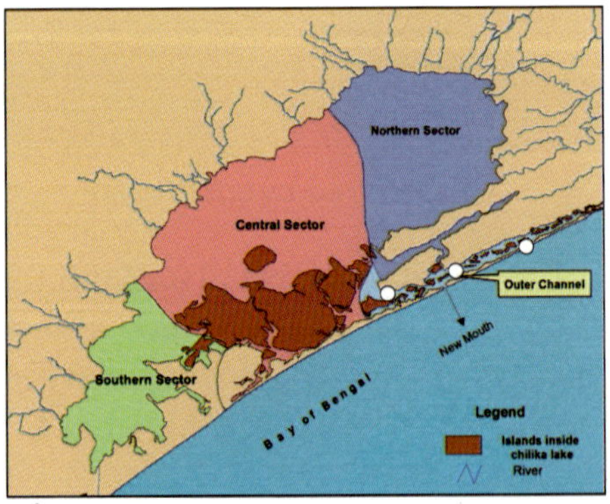

Distribution in Chilika lagoon

Description: Shell thin, elongate ovate, umbo shifted far beyond posterior part; keel absent; posteroventral margin rounded.

Body Colour: White and dark brown.

Distribution in Chilika : In Chilika lake this species is recently recorded from new inlet region, Shipakuda region, Bhabkundaleswar, Old inlet region (Arakhkuda). Earlier it was recorded from Palur Canal, Samal Island, Rambha bay, Barhampur Island, Manikpatna.

Distribution in India: This species is also recorded from **Orissa**: False point; Chandipur, Dhamra estuary; Paradwip, Mahanadi Estuary, Konark; Puri; Chilika lake, Ruashikulya estuary, Chhatrapur, **West Bengal**: Diamond Harbour; Sundarbans,; Digha Coast; Ganga Sagar; Port Canning; Medinipur; North 24 Parganas; South 24 Parganas; Sankarpur; Old Digha, **Gujarat**, **Kerala**,, Maharashtra: Malvan; Ratnagiri; Mumbai; Mahim Creek; Alibag; Nariman Point; Kalbadevi Estuary; Bhatia Creek; Tarkarli Creek, **Goa**: Siridao beach; Siolim; Ribandar; Banastrim; Ribandar; Mandovi Estuary, **Karnataka**: North Kanara; Karwar; Kodibag; Ankola; Tadri, **Pondicherry**: Pondicherry mangrove; **Tamil Nadu**: Parangipettai Coast; Gulf of Mannar; Arukkattuthurai; Kodikkarai; Mallipattinam; Manalmelkudi; Adirampattinam; Adyar Estuary; Vellar Estuary; Cooum Estuary; Pulicat Lake; Pamban Kerala; Porto Novo, Tellicherry; Karangadu; Athankarai; Ennore; Thondi; Karangadu , **Andaman** and **Nicobar Islands**.

Remark: Generally used in decorative purposes. Commonly seen in summer season in the intertidal region of the outer channel area. Dead shells were scattered in the beach sands.

Donax scortum

Phylum : Mollusca

Class: Bivalvia

Order :Veneroida

Family : Donacidae

Genus and species : *Donax scortum*

Donax scortum

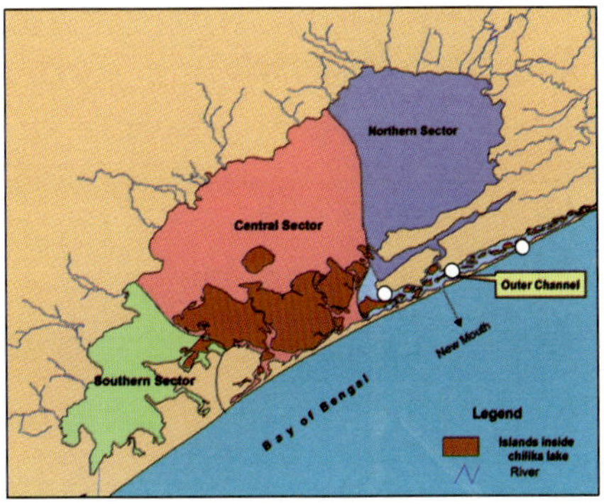

Distribution in Chilika lagoon

Description Size: Length 23.24-79.4 mm; height 15.95-46.34 mm; width 11-34.15 mm. Shell thick, trigonal, inflated; umbo nearly in middle; surface strongly sculptured; prominent keel extends from umbo to postero-ventral margin, truncated.

Body Colour: Dirty white, violet in the posterior region.

Distribution in Chilika : Recent studies in Chilika lake showed that this species is recorded from Shipakuda region, new inlet area, Bhabkundaleswar region. Earlier it was recorded from Palur Canal, Samal Island, Rambha bay, Barhampur Island, Manikpatna , Dead shells were also seen in Balugaon & Kaluparaghat.

Distribution in India: This species is also recorded from **Orissa**: False point; Chandipur, Dhamra estuary; Paradwip, Mahanadi Estuary, Konark; Puri; Chilika lake, Ruashikulya estuary, Chhatrapur, **West Bengal**: Diamond Harbour; Sundarbans,; Digha Coast; Ganga Sagar; Port Canning; Medinipur; North 24 Parganas; South 24 Parganas; Sankarpur; Old Digha, **Gujarat**, **Kerala**,, **Maharashtra**: Malvan; Ratnagiri; Mumbai; Mahim Creek; Alibag; Nariman Point; Kalbadevi Estuary; Bhatia Creek; Tarkarli Creek, **Goa**: Siridao beach; Siolim; Ribandar; Banastrim; Ribandar; Mandovi Estuary, **Karnataka**: North Kanara; Karwar; Kodibag; Ankola; Tadri, **Pondicherry**: Pondicherry mangrove; **Tamil Nadu**: Parangipettai Coast; Gulf of Mannar; Arukkattuthurai; Kodikkarai; Mallipattinam; Manalmelkudi; Adirampattinam; Adyar Estuary; Vellar Estuary; Cooum Estuary; Pulicat Lake; Pamban Kerala; Porto Novo, Tellicherry; Karangadu; Athankarai; Ennore; Thondi; Karangadu , **Andaman and Nicobar Islands**.

Remark: Generally used in decorative purposes. Commonly seen in summer season in the intertidal region of the outer channel area.

Felania Ovalis

Phylum : Mollusca

Class: Bivalvia

Order :Veneroida

Family : Ungulinidae

Genus and species : *Felania ovalis* Preston

Felania annadalei Preston

Diplodonta sataparensis Preston

Diplodonta barhampurensis Preston

Felania ovalis

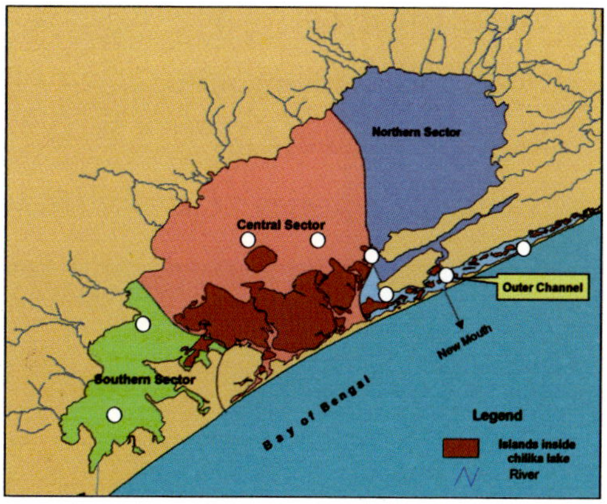

Distribution in Chilika lagoon

Description: *F. ovalis* Shell small, not exceeding 5 mm in length, obliquely ovate. Size: 2.70-12.76 mm

- *D.barhampurensis* - Shell trigonal; umbo large; dorsal margin angularly arched, anterior margin not produced .,Size Length 9.56-23.72 mm; height 7.95-20.90 mm; thickness 4.02-15.00 mm
- *D. satparaensis* - Shell rhomboidal;umbo small, depressed; dorsal margin not arched, anterior margin produced. Size Length 14.50 mm; height 13.70 mm; thickness 10.10 mm
- *F. chilkaensis* - Umbo elevated, posterior margin suddenly slopes, rounded; umbonal area light brown; outer surface olive 'brown. Size: Length 9.90-13.94 mm; height 6.64-13.90 mm; thickness 3.28-6.70 mm
- *F. annandalei* -Umbo not elevated; posterior margin gradually slopes, making angle with ventral margin; outer surface as well as interior white. Size Length 2.70-12.76 mm; height 2.00-11.42 mm; thickness 1.65-5.60 mm

Body Colour: Pale white with brown patches.

Distribution in Chilika : In Chilika lake this species is recorded from Plaur outfall point, Pathara, Nalabana Is. Magarmukh, new inlet region, Old inlet region (Arakhkuda) . Earlier it was recorded from Palur Canal, Samal Island, Rambha bay, Barhampur Island, Manikpatna , Balugaon , Kaluparaghat.

Distribution in India: Endemic to Odisha

Lamellidens marginalis

Phylum : Mollusca

Class: Bivalvia

Order : Pholadomyoida

Family : Pholadidae

Genus and species : *Lamellidens marginalis* Lamarck, 1819

Lamellidens marginalis

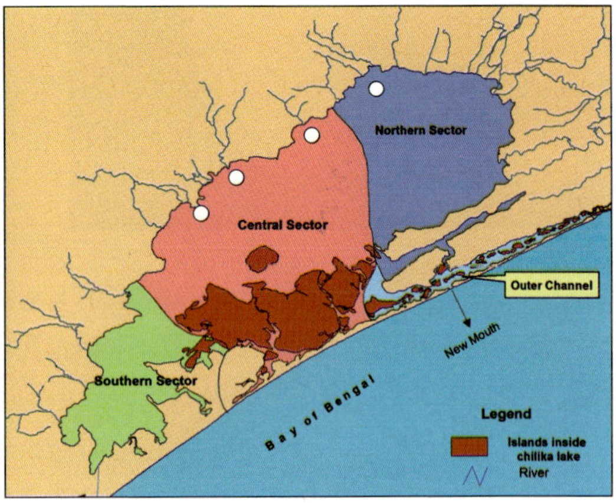

Distribution in Chilika lagoon

Description: Umbonal cavity deep; left valve with two cardinals and two laterals, teeth lamellar, right valve with single cardinal and lateral; interior nacreous, 'outer surface covered with thick periostracum .

Body Colour : Brown to pale brown with a whitish patch at the umbo region.

Distribution in Chilika : Fresh water dominated regions of northern sector and central sectors, river mouth regions.

Distribution in India: Bihar, Haryana, Jharkand, Madhya Pradesh, Rajasthan, Uttaranchal, Uttar Pradesh, West Bengal.

Abroad: Bangladesh; Myanmar; Nepal; Sri Lanka.

Remark: It is a common freshwater bivalve species used as food in many parts of Nepal, India and Bangladesh. It is also used in fresh water pearl culture. Thus over exploitation for feeding and pearl culture, habitat loss, fish poisoning inferred threats to it. In Chilika it embeds in the river mouth bed in the Kalupadaghat region, in Sorana, Tinimuhani, Bhusandapur region etc. It provides a great socio-economic support to the local fishermen since they used it as food as the food particles for prawn culture and shells are used for the lime formation.

Macoma birmanica

Phylum : Mollusca

Class: Bivalvia

Order :Veneroida

Family:Tellinidae

Genus and species: *Macoma birmanica*

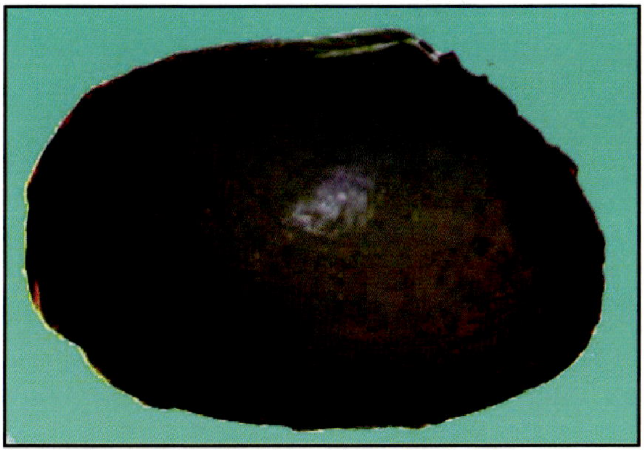

Macoma birmanica

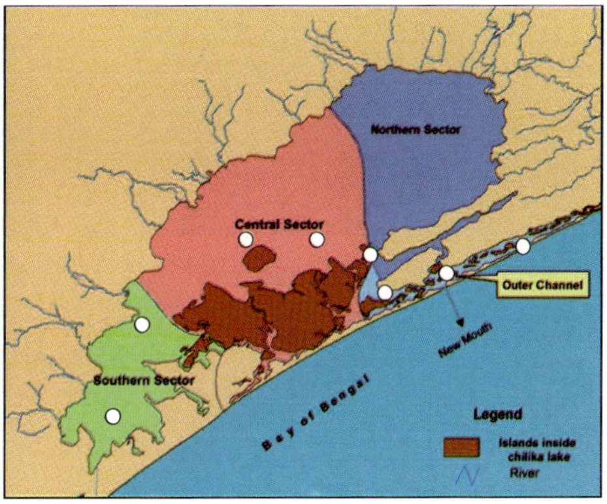

Distribution in Chilika lagoon

Diagnosis: Shell elongately ovate; pallial sinus large occupies entire shell reaching anterior muscle scar.

Body Colour: Ple white to yellowish.

Distribution in Chilika : Collected from the marine influenced region of Chilka lake ranging from the outer channel area, central sector up to the southern sector.

Distribution in India: From Odisha it was previously collected from Talsari, West Bengal, Digha; Chemaguri, Port Canning, Andhra Pradesh: Kakinada Bay.

Remark: This species is abundantly found in the Chilika lake throughout the year. Believed as an important food source for the birds' fishes and shell fishes.

Martesia striata

Phylum : Mollusca

Class: Bivalvia

Order : Myoida

Family : Pholadidae

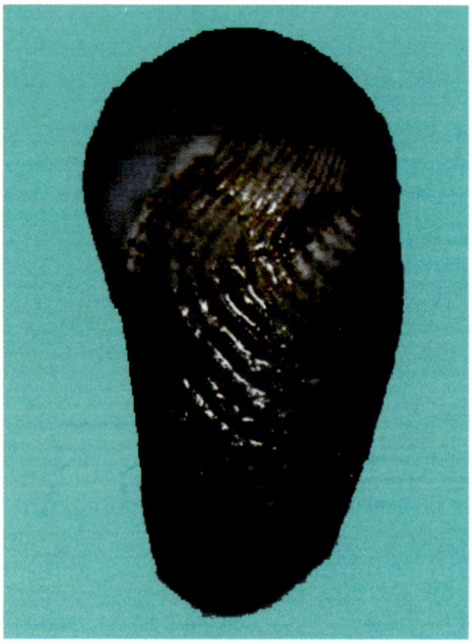

Martesia striata

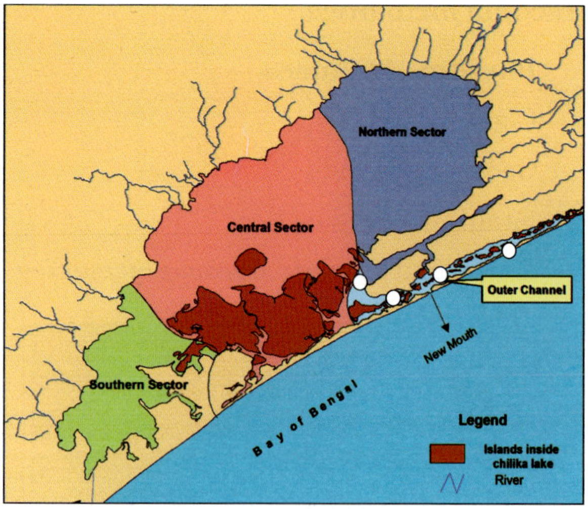

Distribution in Chilika lagoon

Description Size: Length 10.10-16.00 mm; height 6.38-8.36 mm. It is shaped like a block with a narrow posterior and rounded at both ends. It is grayish with equal valves. Siphons are long but fused at the neck.

Body Colour: White with brown rays.

Distribution in Chilika lake : Old mouth (Arakhkuda), New mouth (Shipakuda), Satapada, Rambharatia is. Magarmukh.

Distribution in India : Andaman and Nicobar Islands, Orissa, Dhamara estuary; Mahanadi Estuary; Chilika Lake, Gopalpur; Chhatrapur; Rushikulya estuary; Hukitola; Chandipur; West Bengal: Jharkhali; North 24 Parganas; Sundarbans; Champta. Andhra Pradesh, Chittivalasa; Gujarat : Gulf of Kachchh, Karnataka, Kerala, Maharashtra, Tamil Nadu : Ennore Backwaters, Gulf of Mannar; Krusadai; Shingle Island, Kerla, Kochi; Kozhikode, Karnataka: Karwar, Maharashtra: Mumbai, Lakshadweep.

Abroad: Circumglobal distribution in temperate and tropical marine waters.

Remark : *M. striata* is one of the most destructive warm-water wood boring organisms . Members of this family bore into submerged wooden structures, soft rocks. These are also distributed in the tropical and subtropical waters. About 7 species under 5 genera, namely *Martesia*, *Pholas*, *Lignopholas*, *Barnea* and *Joullnetia* are known from India. The first mentioned is the well known genus and has three species in India. *M. striata* is the most common species occurring in estuaries of India. Judged from its habitat requirement, the species may be able to survive in the Lake, especially in its outer channel.

Meretrix casta

Phylum : Mollusca

Class: Bivalvia

Order :Veneroida

Family:Veneridae

Genus and species: *Meretrix casta* Gmelin

Meretrix casta

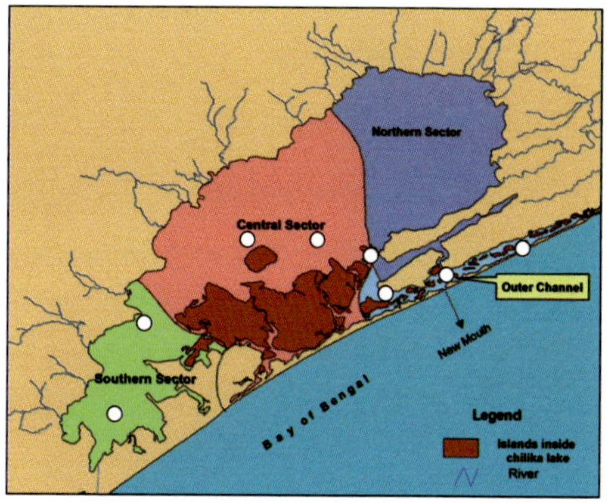

Distribution in Chilika lagoon

Description Size: Length 12.50-55.95 mm; height 10.80-53.58 mm. Hinge broad, shell higher than long; valves thick, more inflated, posterior end without band. This clam has a thick, moderately large shell with a brown horny periostracum. A dark greyish band is present at the posterior margin of the shell. The shell is ovate and heavy.

Distribution in Chilika : In Chilika lake this species is recorded from Plaur outfall point, Pathara, Nalabana Is. Magarmukh, new inlet region, Old inlet region (Arakhkuda) . Earlier it was recorded from Palur Canal, Samal Island, Rambha bay, Barhampur Island, Manikpatna.

Distribution in India: This species is also recorded from **Orissa**: False point; Chandipur, Dhamra estuary; Paradwip, Mahanadi Estuary, Konark; Puri; Chilika lake, Ruashikulya estuary, Chhatrapur, **West Bengal**: Diamond Harbour; Sundarbans,; Digha Coast; Ganga Sagar; Port Canning; Medinipur; North 24 Parganas; South 24 Parganas; Sankarpur; Old Digha, **Gujarat**, **Kerala**,, **Maharashtra**: Malvan; Ratnagiri; Mumbai; Mahim Creek; Alibag; Nariman Point; Kalbadevi Estuary; Bhatia Creek; Tarkarli Creek, **Goa**: Siridao beach; Siolim; Ribandar; Banastrim; Ribandar; Mandovi Estuary, **Karnataka**: North Kanara; Karwar; Kodibag; Ankola; Tadri, **Pondicherry**: Pondicherry mangrove; **Tamil Nadu**: Parangipettai Coast;

Gulf of Mannar; Arukkattuthurai; Kodikkarai; Mallipattinam; Manalmelkudi; Adirampattinam; Adyar Estuary; Vellar Estuary; Cooum Estuary; Pulicat Lake; Pamban Kerala; Porto Novo, Tellicherry; Karangadu; Athankarai; Ennore; Thondi; Karangadu , **Andaman and Nicobar Islands**.

Remark: Acquire high antiviral activity; used for making lime, Edible.

Meretrix meretrix

Phylum : Mollusca

Class: Bivalvia

Order : Veneroida

Family: Veneridae

Genus and species: *Meretrix meretrix* Linnaeus, 1758

Meretrix meretrix

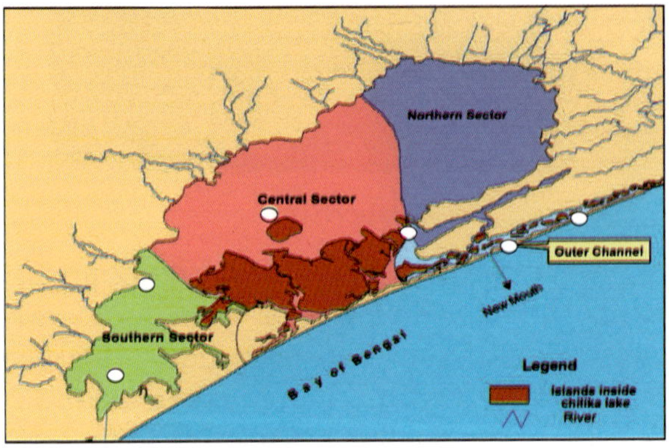

Distribution in Chilika lagoon

Description Size: Maximum shell length 6. cm; width 8-60 mm; height 4.1-40.5 mm.Shell thick, moderately inflated with a variable, nearly equilateral to rather strongly inequilateral shape, trigonal-ovate in outline. Umbones anterior, poorly inflated, pointing on top of dorsal margin which is strongly sloping on both sides and longest posteriorly. Anterior and ventral margins broadly rounded, posterior end of shell bluntly angled. Lunule smooth and poorly defined, escutcheon indistinct. Posterodorsal slope shallowly convex, bordered by a faint radial fold that disappears toward posteroventral end of valves. Outer surface of shell smooth, except from low concentric growth marks. Periostracum smooth and glossy, closely applied to shell. Irregularly shaped hinge plate thick. On the postero-dorsal margin of the shell a greyish blue or bluish brown band is present.

Body color: Colour of the shell showed uneven color pattern, pale straw-colored periostracum, with deep purple brown on posterodorsal slope, blotched or streaked with darker grey or brown. Interior porcelaneous white, sometimes stained with dark brown along posterodorsal margin.

Distribution in Chilika : In Chilika lake this species is recorded from Plaur outfall point, Pathara, Nalabana Is. Magarmukh, new inlet region, Old inlet region (Arakhkuda) . Earlier this species is observed from the locations such as Satpada, Arakhkuda, Pathara and Palur canal.

Distribution in India: This species is also recorded from **Orissa**: Subarnarekha Mouth; Talasari, Chandipur, Dhamra estuary; Paradwip, Mahanadi Estuary, Konark; Chilika lake, Ruashikulya estuary, Chhatrapur, **West Bengal**: Diamond Harbour; Sundarbans,; Digha Coast; Ganga Sagar; Port Canning; Medinipur; North 24 Parganas; South 24 Parganas; Sankarpur; Old Digha; **Andaman and Nicobar Islands**, **Gujarat**, **Kerala**,, **Maharashtra**: Malvan; Ratnagiri; Mumbai; Mahim Creek; Alibag; Nariman Point; Kalbadevi Estuary; Bhatia Creek; Tarkarli Creek, Goa: Siridao beach, Siolim, Ribandar, Karnataka: North Kanara; Karwar; Kodibag; Ankola; Tadri, **Pondicherry**: Pondicherry mangrove; **Tamil Nadu**: Parangipettai Coast; Gulf of Mannar; Arukkattuthurai; Kodikkarai; Mallipattinam; Manalmelkudi; Adirampattinam; Adyar Estuary; Vellar Estuary; Cooum Estuary; Pulicat Lake; Pamban Kerala; Porto Novo, Tellicherry.

Abroad: Red Sea , Sri Lanka, Myanmar.

Modiolus modiolus

Phylum : Mollusca

Class: Bivalvia

Order : Mytiloida

Family : Mytilidae

Genus and species : *Brachidontes striatulus* Hanley, 1843

Brachidontes undulates Dunker, 1857

Modiolus modiolus Linnaeus, 1758

Brachidontes modiolus Linnaeus, 1767

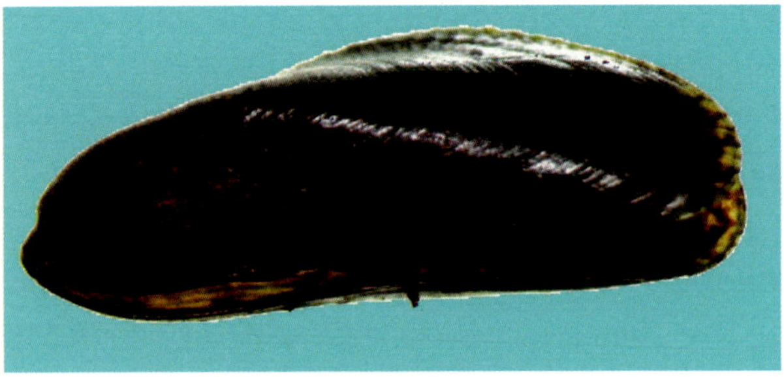

Modiolus modiolus

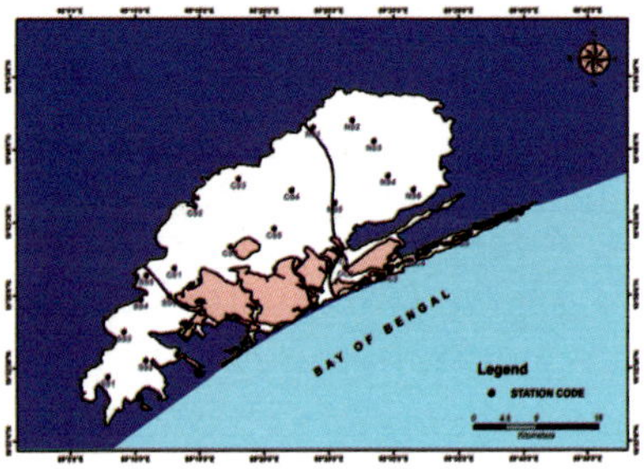

Distribution in Chilika lagoon

Description Size: Length 4.95-22.58 mm; height 3.00-9.25 mm; thickness 1.82-8.15 mm.Upper margin strongly elevated and evenly arched, postero-dorsal margin declines and posterior extremity narrowed. Shells are usually found attached to rocks, stones, wooden jetties in submerged waters, and also on algal growths on stone.

- ***Modiolus modiolus*** **-** Shell thin, sometimes transparent; small, not exceeding 30mm in length; umbo situated behind anterior end; surface not covered with periostracum, radially striated.
- ***Modiolus striatulus*** - Ventral margin concave in middle, dorsal margin evenly arched; surface without undulating markings
- ***Modiolus undulates*-** Ventral margin straight; dorsal margin not arched; surface with pinkish brown undulating markings

Distribution in India: This species is coomonly recorded from the east coast and west coast of India. However some of the conspicuous ecosystems are those of **Orissa**: False point; Chandipur; Dhamra estuary; Paradwip; Mahanadi Estuary; Konark; Puri; Chilika lake; Ruashikulya estuary; Chhatrapur; Gopalpur, **West Bengal**: Diamond Harbour; Sundarbans,; Digha Coast; Ganga Sagar; Port Canning; Medinipur; North 24 Parganas; South 24 Parganas; Sankarpur; Old Digha, **Gujarat**, **Kerala**,, **Maharashtra**: Malvan; Ratnagiri; Mumbai; Mahim Creek; Alibag; Nariman Point; Kalbadevi Estuary; Bhatia Creek; Tarkarli Creek, **Goa**: Siridao beach; Siolim; Ribandar; Banastrim; Ribandar; Mandovi Estuary, **Karnataka**: North Kanara; Karwar; Kodibag; Ankola; Tadri, **Pondicherry**: Pondicherry mangrove; **Tamil Nadu**: Parangipettai Coast; Gulf of Mannar; Arukkattuthurai; Kodikkarai; Mallipattinam; Manalmelkudi; Adirampattinam; Adyar Estuary; Vellar Estuary; Cooum Estuary; Pulicat Lake; Pamban Kerala; Porto Novo, Tellicherry; Karangadu; Athankarai; Ennore; Thondi; Karangadu , **Andaman and Nicobar Islands**.

Remark: This found all most all the region of Chilika: in the pillars, sea grass, sea weeds, sponges, macrophytes . Found all most in all the seasons. Small juveniles are produced during june and July and they get young size during November and December and adult size during march and April. Important food for crabs, prawns even for the migratory birds. In the Chilika lake this species is accommodated with all the physic-chemical heterogeneity.

Perna viridis

Phylum : Mollusca

Class: Bivalvia

Order : Mytiloida

Family : Mytilidae

Genus and species : *Perna viridis* Linnaeus, 1758

Perna viridis

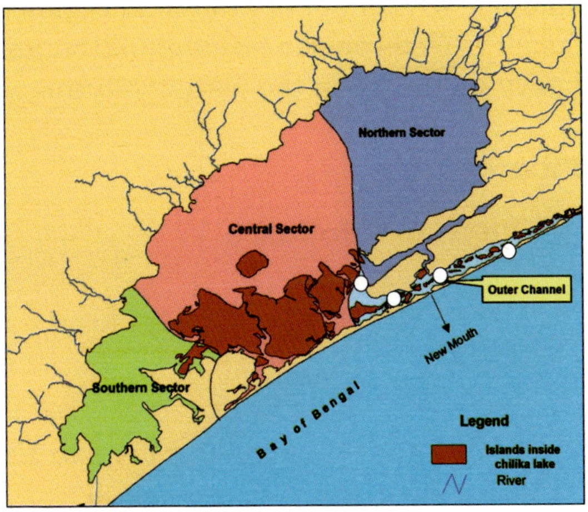

Distribution in Chilika lagoon

Description Size: Length 80-100 mm, occasionally reaching 165 mm; width 5.50 mm; height 2.85 mm.Shell thick, large, exceeds 30nlm in length; umbo situated at anterior end; surface covered with brown periostracum; green in colour; concentrically striated .

Body colour: Green brown in colour.

Distribution in Chilika : In Chilika lake this species is recorded from Plaur outfall point, Pathara, Nalabana Is. Magarmukh, new inlet region, Old inlet region (Arakhkuda) . Earlier it was recorded from Palur Canal, Samal Island, Rambha bay, Barhampur Island, Manikpatna.

Distribution in India: This species is also recorded from **Orissa**: False point; Chandipur, Dhamra estuary; Paradwip, Mahanadi Estuary, Konark; Puri; Chilika lake, Ruashikulya estuary, Chhatrapur, **West Bengal**: Diamond Harbour; Sundarbans,; Digha Coast; Ganga Sagar; Port Canning; Medinipur; North 24 Parganas; South 24 Parganas; Sankarpur; Old Digha, **Gujarat**, **Kerala**,, **Maharashtra**: Malvan; Ratnagiri; Mumbai; Mahim Creek; Alibag; Nariman Point; Kalbadevi Estuary; Bhatia Creek; Tarkarli Creek, **Goa**: Siridao beach; Siolim; Ribandar; Banastrim; Ribandar; Mandovi Estuary, **Karnataka**: North Kanara; Karwar; Kodibag; Ankola; Tadri, **Pondicherry**: Pondicherry mangrove; **Tamil Nadu**: Parangipettai Coast; Gulf of Mannar; Arukkattuthurai; Kodikkarai; Mallipattinam; Manalmelkudi; Adirampattinam; Adyar Estuary; Vellar Estuary; Cooum Estuary; Pulicat Lake; Pamban Kerala; Porto Novo, Tellicherry; Karangadu; Athankarai; Ennore; Thondi; Karangadu , **Andaman and Nicobar Islands**.

Remark: Generally used in decorative purposes. Commonly seen in summer season in the intertidal region of the outer channel area. Being a filter feeder it feeds on the phytoplankton and also accumulates virus, bacteria and heavy metals from the water column.

Pitar erycina

Phylum : Mollusca

Class: Bivalvia

Order :Veneroida

Family:Veneridae

Genus and species: *Pitar erycina* Linnaeus, 1758

Pitar erycina

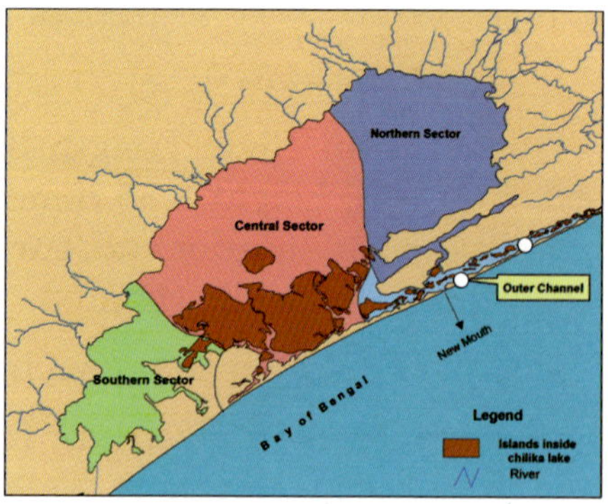

Distribution in Chilika lagoon

Description: The shell is thicker and the umbo is inclined forwards. The shell length is greater than the height.Strong concentric grooves with moderately broad flattened spacefarer seen on the surface of the shell. The hinge teeth are solid, the infront tooth of the cardinals well developed and posterior lateral tooth narrow and much elongated. Outer surface is pale brown, touched with orange red towards the anterior and ventral margins. Noticeably rayed radially, chocolate brown bands of different widths deviate from the umbo. Region behind the umbo directly surrounding the lunule are traversed by widely spaced brown strips.

Body Colour: Yellowish brown in colour.

Distribution in Chilika lake: First time recorded from Chilika lake during the year 2007 to 2010 integrated macrobenthos survey . Rarely reported from the outer channel area mostly from the Bhavkundaleswar region (OC4) and Old mouth-Arakhkuda region (OC5) during late winter and summer seasons.

Distribution in India: Tamil Nadu, Parangipettai Coast; Mandapam; Dhanushkodi, Goa, Marmagao.

Remark: Dead shells are collected for lime preparation. Live shells are found in the high saline phase in the lake. Strictly confined to the outer channel area. It is believed that after opening of the new lagoonal inlet. This species might have invade the lake area during high tide periods and gradually accommodated with the lake condition.

Placuna placenta

Phylum : Mollusca

Class: Bivalvia

Order : Pectinoida

Family : Anomiidae

Genus and species: *Placuna placenta*

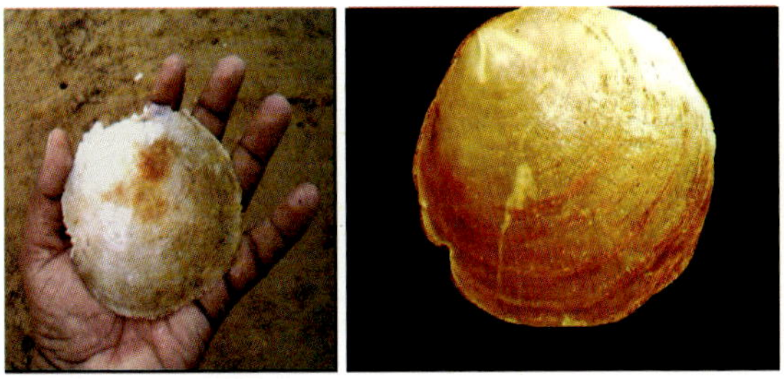

Placuna placenta

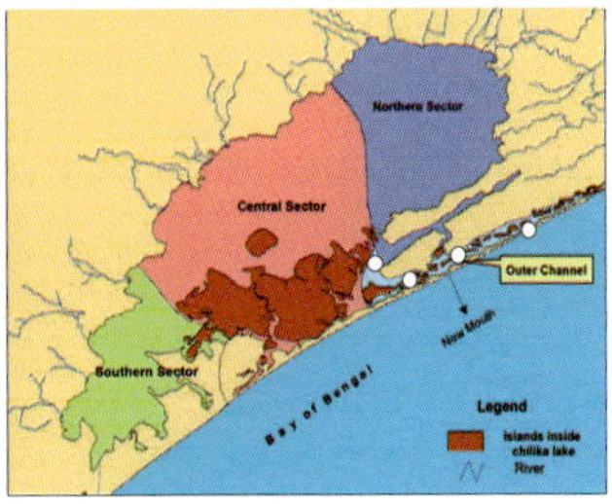

Distribution in Chilika lagoon

Description: Shell orbicular, large, exceeding 40mm in length, compressed, upper valve slightly inflated, lower valve flat; without hole. Size: Maximum shell length 18 cm, commonly to 10 cm. **Body Color**: Outside of shell silvery white with a dull finish, occasionally with pale brown or light purplish rays umbonally. Interior is nacreous white.

Shell is thin and more or less translucent, almost circular in outline. Dorsal margin somewhat flattened to widely curved, sometimes faintly protruding anteriorly and posteriorly. Valves greatly compressed laterally, the lower (right) valve flat, the upper (left) valve with slight convexity. Commissural plane is flat (rarely, if at all, subflexuous). Outer surface is nearly smooth, excepting numerous, minute radiating threads forming tenuous wrinkles on the finely lamellate concentric lines of growth.

Distribution in Chilika lake : Old mouth (Arakhkuda), New mouth (Shipakuda), Satapada, Rambharatia is.

Distribution in India : Andaman and Nicobar Islands, Orissa, Dhamara estuary; Mahanadi Estuary; Chilika Lake, Gopalpur; Chhatrapur; Rushikulya estuary; Hukitola; Chandipur; West Bengal: Jharkhali; North 24 Parganas; Sundarbans; Champta. Andhra Pradesh, Chittivalasa; Gujarat : Gulf of Kachchh, Karnataka, Kerala, Maharashtra, Tamil Nadu : Ennore Backwaters, Gulf of Mannar; Krusadai; Shingle Island, Kerla, Kochi; Kozhikode, Karnataka: Karwar, Maharashtra: Mumbai, **Lakshadweep**.

Remark: Dead shells are found through out the year while live shells are found in summer seasons only. Often observed in the lowtide periods. Their no. is found higher after the opening of the lagoonal inlet.

Saccostrea cucullata

Phylum : Mollusca

Class: Bivalvia

Order : Ostreoida

Family : Ostreidae

Genus and species: *Saccostrea cucullata* Born 1778

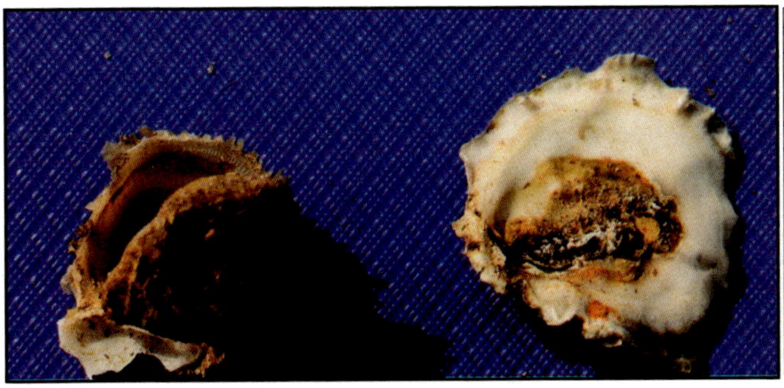

Saccostrea cucullata

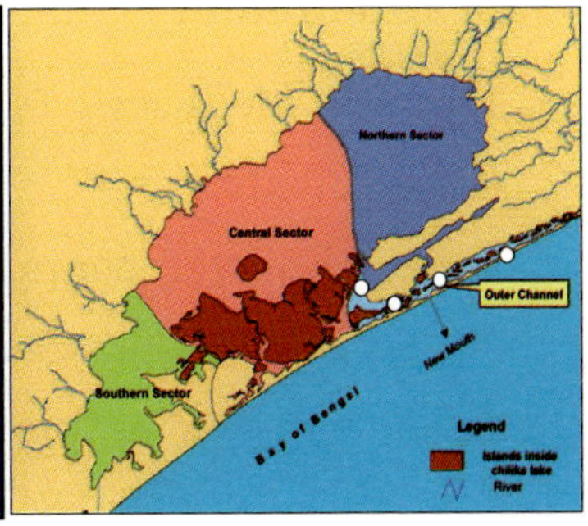

Distribution in Chilika lagoon

Description Size: Length 20.60-29.80 mm; width 44.70-56.80 mm; height 14.40-18.40 mm. Shell small not exceeding 50mm in length; lower valve with its margin crenulate; rows of tubercles present on either side of the umbonal grove.

Body colour: Deep violet. Adductor muscles white in color.

Distribution in Chilika : Outer channel area only.

Distribution in India: This species is also recorded from **Orissa**: False point; Chandipur, Dhamra estuary; Paradwip, Mahanadi Estuary, Konark; Puri; Chilika lake, Ruashikulya estuary, Chhatrapur, **West Bengal**: Diamond Harbour; Sundarbans,; Digha Coast; Ganga Sagar; Port Canning; Medinipur; North 24 Parganas; South 24 Parganas; Sankarpur; Old Digha, **Gujarat**, **Kerala**,, Maharashtra: Malvan; Ratnagiri; Mumbai; Mahim Creek; Alibag; Nariman Point; Kalbadevi Estuary; Bhatia Creek; Tarkarli Creek, **Goa**: Siridao beach; Siolim; Ribandar; Banastrim; Ribandar; Mandovi Estuary, **Karnataka**: North Kanara; Karwar; Kodibag; Ankola; Tadri, **Pondicherry**: Pondicherry mangrove; **Tamil Nadu**: Parangipettai Coast; Gulf of Mannar; Arukkattuthurai; Kodikkarai; Mallipattinam; Manalmelkudi; Adirampattinam; Adyar Estuary; Vellar Estuary; Cooum Estuary; Pulicat Lake; Pamban Kerala; Porto Novo, Tellicherry; Karangadu; Athankarai; Ennore; Thondi; Karangadu , **Andaman and Nicobar Islands**.

Remark: The hooded oyster cements itself to rocks, branches and roots of mangroves, or any hard structures. Occasionally grows well on the shell of a large gastropod mollusk. It is a filter feeders feeds on phytoplankton and detritus. It also accumulates microbes and heavy metals on their body. In Chilika it is mostly seen during summer season and found scattered in the low tide periods.

Solen annandalei

Phylum : Mollusca

Class: Bivalvia

Order : Euheterodonta

Family : Solenidae

Genus and species : *Solen annandalei* Preston, 1915

Solen kempi Preston

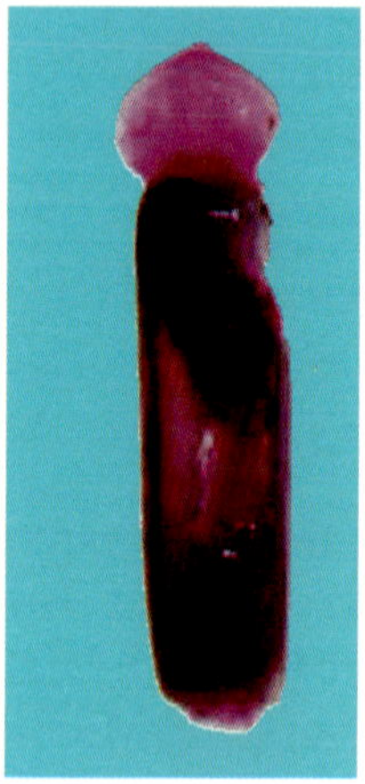

Solen Kempi

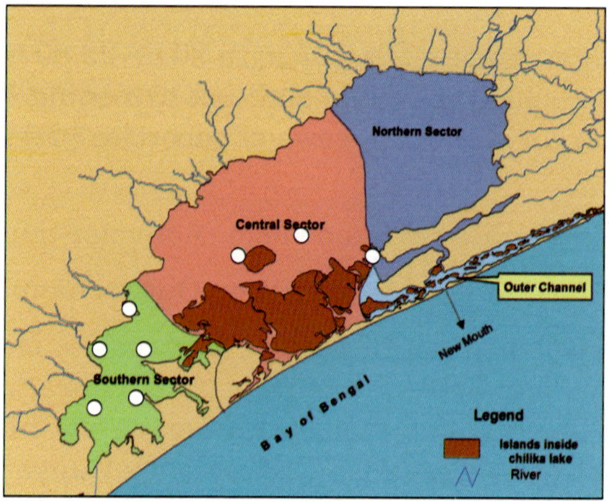

Distribution in Chilika lake

Description Size: Length 31.15-77.5 mm; height 6.10-14.25 mm; thickness 4.22-9.90 mm.Shell broader, posterior part equal to anterior; posterodorsal margin rounded.

- ***Solen kempi:*** Shell linear, narrow, posterior part narrower than anterior; posterodorsal margin angular.
- ***S.annandalei:*** Shell broader, posterior part equal to anterior; posterodorsal margin rounded.

Distribution in Chilika : Southern sector, Central sector and outer Magarmukh area.

Distribution in India: Odisha, Chilika lake, Rushkikulya estuary, Dhamra estuary, Tamilnadu coast and Kerla coast.

Remark : A large family of burrowing forms with a worldwide distribution. The genus *Solen* with 6 species are recorded from India. These were recorded from intertidal mudflats or sandy beaches. Two species namely, *Solen* *annandalei* Preston and *S. kenlpi* Preston were originally described from the lake. *Solen annandaleii* is widely recorded in the southern sector and few region of central sector of Chilika lake.

Sunetta scripta

Phylum : Mollusca

Class: Bivalvia

Order :Veneroida

Family: Veneridae

Genus and species: *Sunetta scripta* Linnaeus

Sunetta scripta

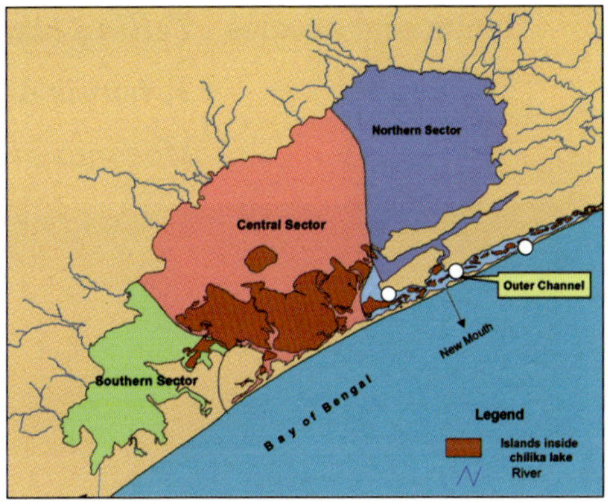

Distribution in Chilika lagoon

Description: Shell is smooth, glossy and polished, moderately elongated, evenly rounded in front and obliquely truncated behind. Shell has flattened and somewhat faint concentric lines. Hinge teeth are thin and narrow and the tooth in front of the cardinals in the left valve and the hollow in the right are considerably elongated. Lunule is long, narrow and deeply excavated. The area behind the umbo is also deeply sunk. Inside of the shell is white and purplish, glossy with moderately deep pallial sinus, deeply sinuated pallial line and crenulated rim.

Body Colour: Outer surface is generally pale yellowish or flashy white and bears a bright color pattern or arrow-head shaped markings.

Distribution in Chilika : During the present study in Chilika this species is distributed in Entire outer channel area. Abundantly found during summer season.

Distribution in India: It is distributed both in the east coast and west coast of India as well along with the Andaman and Nicobar Islands.

Abroad: Circum-tropical.

Remarks : Commonly found during the low tide period.

Tellina texturata

Phylum : Mollusca

Class: Bivalvia

Order :Veneroida

Family : Tellinidae

Genus and species : *Tellina texturata* Sowerby in Reeve's 1867

T. tenuis da Costa, 1778

Macoma birmanica Philippi

Tellina texturata

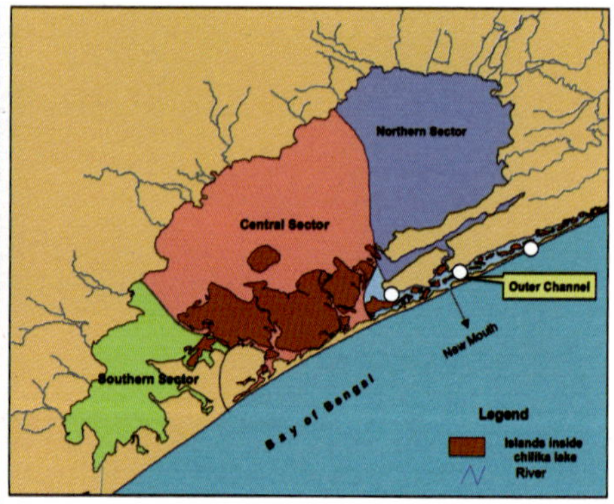

Distribution in Chilika lagoon

Description Size: Length 7.00-9.70 mm; height 4.26-6.20 mm; thickness 2.13-3.60, Shell elongate ovate; umbonal region white; surface not shining.

Distribution in Chilika : In Chilika lake this species is recorded from Plaur outfall point, Pathara, Nalabana Is. Magarmukh, new inlet region, Old inlet region (Arakhkuda) . Earlier this species is observed from the locations such as Satpada, Arakhkuda, Pathara and Palur canal.

Distribution in India: This species is also recorded from Orissa: Subarnarekha Mouth; Talasari, Chandipur, Dhamra estuary; Paradwip, Mahanadi Estuary, Konark; Chilika lake, Ruashikulya estuary, Chhatrapur,

West Bengal: Diamond Harbour; Sundarbans,; Digha Coast; Ganga Sagar; Port Canning; Medinipur; North 24 Parganas; South 24 Parganas; Sankarpur; Old Digha,; Andaman and Nicobar Islands, **Gujarat**, **Kerala**,, **Maharashtra**: Malvan; Ratnagiri; Mumbai; Mahim Creek; Alibag; Nariman Point; Kalbadevi Estuary; Bhatia Creek; Tarkarli Creek, Goa: Siridao beach, Siolim, Ribandar, **Karnataka**: North Kanara; Karwar; Kodibag; Ankola; Tadri, Pondicherry: Pondicherry mangrove; **Tamil Nadu**: Parangipettai Coast; Gulf of Mannar; Arukkattuthurai; Kodikkarai; Mallipattinam; Manalmelkudi; Adirampattinam; Adyar Estuary; Vellar Estuary; Cooum Estuary; Pulicat Lake; Pamban Kerala; Porto Novo, Tellicherry.

Abroad: Red Sea , Sri Lanka , Myanmar.

Tegillarca granosa

Phylum : Mollusca

Class: Bivalvia

Order : Arcoida

Family : Arcidae

Genus and species : *Striarca lacteal* Linnaeus, 1758
Tegillarca granosa Linnaeus, 1758
Tegillarca rhombea Born, 1778
Scapharca deyrollei Jousseaume, 1893

Tegillarca granosa

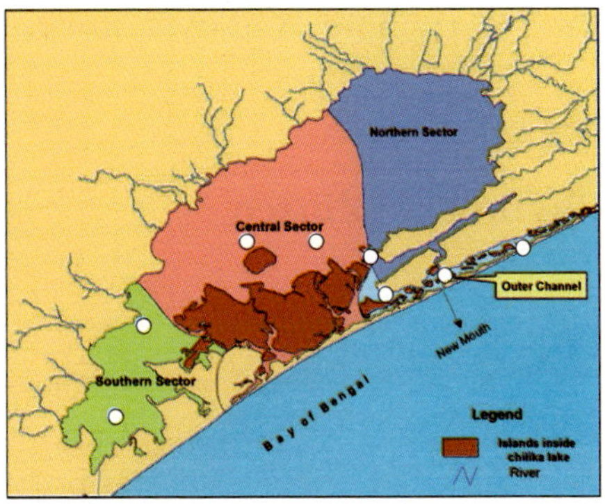

Distribution in Chilika lagoon

Description: Shell thick but not rhomboidal, more elongate; keel on posterior part absent, ribs nodulose, equivalve; surface sculptured with nodulose radial ribs.

Body Colour: white, pale brown in colour.

Distribution in Chilika : In Chilika lake this species is recorded from Plaur outfall point, Pathara, Nalabana Is. Magarmukh, new inlet region, Old inlet region (Arakhkuda) . Earlier it was recorded from Palur Canal, Samal Island, Rambha bay, Barhampur Island, Manikpatna , Balugaon , Kaluparaghat ,

Distribution in India: This species is also recorded from **Orissa**: False point; Chandipur, Dhamra estuary; Paradwip, Mahanadi Estuary, Konark; Puri; Chilika lake, Ruashikulya estuary, Chhatrapur, **West Bengal**: Diamond Harbour; Sundarbans,; Digha Coast; Ganga Sagar; Port Canning; Medinipur; North 24 Parganas; South 24 Parganas; Sankarpur; Old Digha, **Gujarat**, **Kerala**,, **Maharashtra**: Malvan; Ratnagiri; Mumbai; Mahim Creek; Alibag; Nariman Point; Kalbadevi Estuary; Bhatia Creek; Tarkarli Creek, **Goa**: Siridao beach; Siolim; Ribandar; Banastrim; Ribandar; Mandovi Estuary, **Karnataka**: North Kanara; Karwar; Kodibag; Ankola; Tadri, **Pondicherry**: Pondicherry mangrove; **Tamil Nadu**: Parangipettai Coast; Gulf of Mannar; Arukkattuthurai; Kodikkarai; Mallipattinam; Manalmelkudi; Adirampattinam is eaten but used as food mat;s meat Adyar Estuary; Vellar Estuary; Cooum Estuary; Pulicat Lake; Pamban Kerala; Porto Novo, Tellicherry; Karangadu; Athankarai; Ennore; Thondi; Karangadu , **Andaman and Nicobar Islands**.

Remark: In Chilika it used for the lime preparation. Its meat is used in prawn hatcheries but human edible reports are less in and around Chiulika. It is also called as blood cockle clam because of the red hemoglobin liquid inside the soft tissues. It ingests more viruses and bacteria, including hepatitis A, hepatitis E, typhoid, and dysentery.

Theora opalina

Phylum : Mollusca

Class: Bivalvia

Order: Veneroida

Family: Semelidae

Genus and species: *Theora opalina* Preston 1914

Cumingia hinduorum Annan. & Kemp, 1916

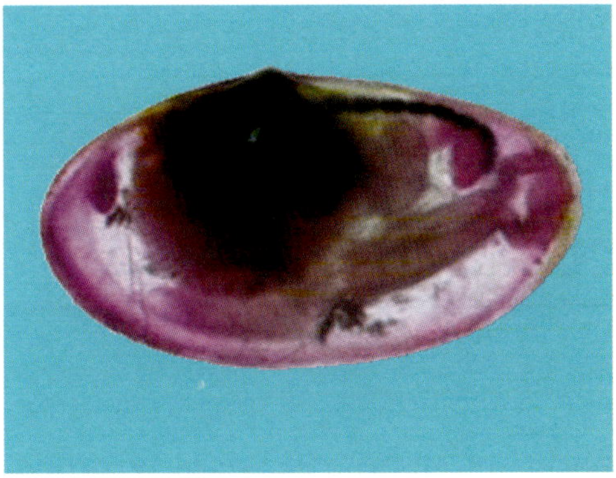

Theora opalina

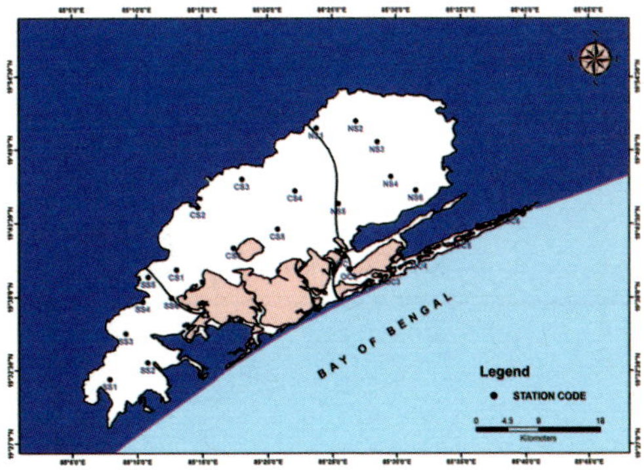

Distribution in Chilika lagoon

Description Size: Length 4.-15.92 mm; width 2.3-8.5 mm; height 1.0-3.1 mm.Shell oval, thin, fragile translucent, smooth, compressed Umbo acuminate and anterior in position. Anterior dorsal straight and gradually sloped. Posterior dorsal straight. Ventral margin convex. Sculpture smooth or with fine concentric posterior part without keel Muscle scars well impressed.

Body Color: Creamy white.

Distribution in Chilika : Nalbana bird Is., Kaluparaghat, Balugaon, Badakuda, Sanakuda, Samanashi, Satapada, Magarmukh, Outer channel area, Throughout the lake.

Distribution in India: Odisha: Puri, Konark, Mahanadi estuary, Rushikulya estuary, West Bengal, Mooriganga Estuary; Sagar Island; Port Canning; Tamil Nadu, Ennore; Chennai, Kerala, Malabar Coast; Kerala, Kochi, Maharashtra, Mumbai.

Remark: This species is found throughout the year and mostly comprosed to the southern and central sector. More fresh water or saline water it commonly not visible. It is preferred food stuff of birds, crabs, and prawns.

6

Gastropods: Pictorial guide

During the present study a total of 82 species of gastropods were recorded. Most of the species are common and recorded previously where as few of the newly occurring species were also recorded during this study. Most of the new records were made from the outer channel area. The taxonomic classification, description, distribution details of certain newly recorded gastropods are given here under.

Aplysia dactylomela

Phylum : Mollusca

Class: Gastropoda

Order : Anaspidea

Family : Aplysiidae

Genus and species : *Aplysia dactylomela* Rang, 1828

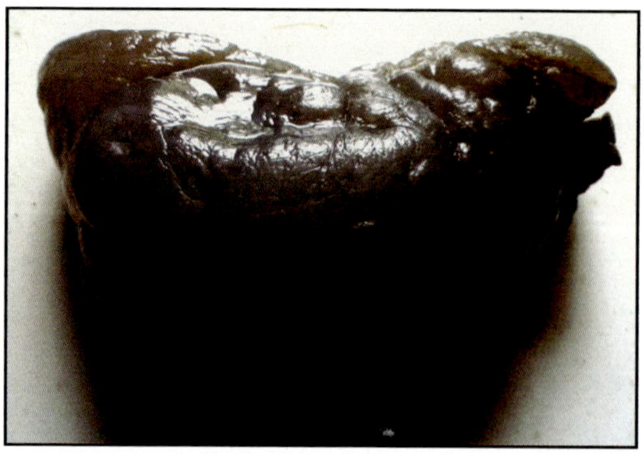

Aplysia dactylomela

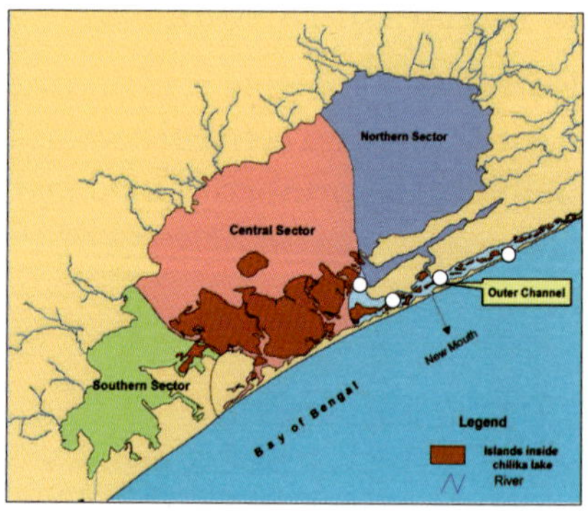

Distribution in Chilika lagoon

Description Size: Max length 40.0 cm TL (male) Characterized by relatively tough, leathery parapodia and large dark or black rings, and blackish reticulate lines. Black markings on both the outside and inside of the parapodia.

Body Color: Color variation with relatively thin black lines and rings.

Distribution in Chilika : In Chilika lake this species is recorded from Magarmukh, new inlet region, Old inlet region (Arakhkuda).

Distribution in India: This species is also recorded from Odisha:Chilika lake, Tamilnadu, Pulicate Lake.

Abroad: Circum-tropical.

Remark: This species is a rare constituent of the sea grass bed of Chilika. It is commonly found in shallow waters, tide pools and rocky and sandy substrates; they also will be found feeding in beds of sea grass. During the day they will mostly hide under large rocks or in the dense sea grass beds. They usually stay in relatively shallow coastal waters. When disturbed *Aplysia dactylomela* releases a purple ink that makes irritation. Their leathery skin contains toxins which make this sea hare practically inedible to most predators.

Armina maculata

Phylum : Mollusca

Class: Gastropoda

Order : Nudibranchia

Family : Arminidae

Genus and species : *Armina maculata* Rafinesque, 1814

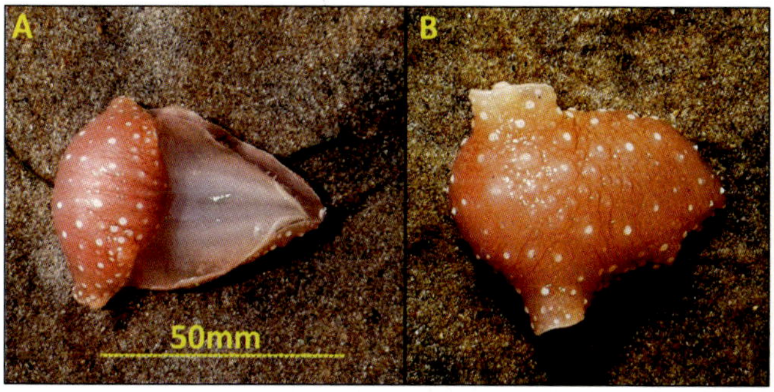

Armina maculata

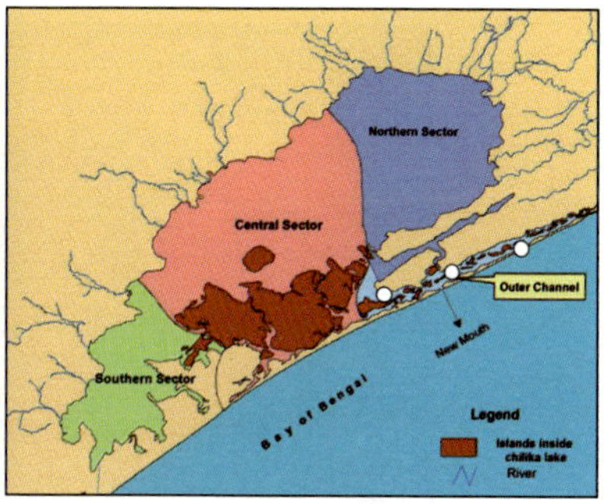

Distribution in Chilika

Description: The body is elongated, flattened in the middle, and gradually narrows towards the posterior. The general body color is orange with white pustules. The foot is white with spherical pustules of different sizes found on the frontal part. These pustules decrease in number towards the medial part of the body and increase towards the posterior. The rhinophores are situated close together, their notal ridges with irregularly arranged pustules.

Body colour: In live animals, the general body color is orange but the pustules and foot are white.

Distribution in Chilika : New record to Chilika. In Chilika lake this species is recorded from Magarmukh, new inlet region, Old inlet region (Arakhkuda).

Distribution in India: Not reported till date.

Abroad: Mediterranean sea, Turkey.

Remark: In India, four species of *Armina* are reported: *Armina cinerea* from Gulf of Kuchh *Armina cygnea* and *Armina semperi* from Andaman waters, and *Armina babai* from Parangipettai coast Tamil Nadu. Some of these species were found in the coral reef bed regions of the Andaman and Nicober Islands of India.

Babylonia spirata

Phylum : Mollusca

Class: Gastropoda

Order : Caenogastropoda

Family : Babyloniidae

Genus and species : *Babylonia spirata* Linnaeus, 1758

Babylonia spirata

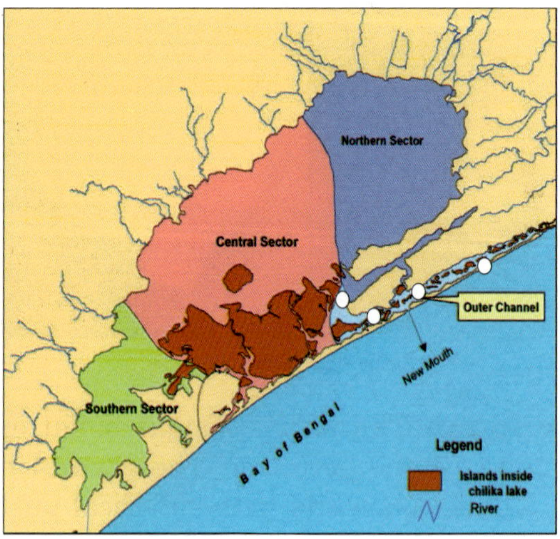

Distribution in Chilika lagoon

Description: Size: Length 25.65-52.96 mm; diameter 17.50-37.30 mm. The shell is smooth, ovoid prominent shoulders, deep incised sutures, turreted like and covered when fresh by a strong brownish periostracum. The whorls are inflated and sutures are sunk in broad, inwardly sloping grooves. The columella is smooth and somewhat strongly thickened with heavy callus; the umbilicus is almost completely filled up by callus. The aperture is large, ovate and constricted posterior by a single, thick ridge extending spirally inverse on the columellar side. The anterior canal is represented by an oblique notch at the base of the aperture. The inside of the outer lip is smooth and unthickened.

Body colour: White with reddish brown blotches; regular spiral rows of large, rounded or Squamish brownish or orange brown patches on a white ground; the apex of the aperture is tinged blackish.

Distribution in Chilika: In Chilika lake this marine gastropod is recorded from outer channel area only. Some dead shells are occupied by the hermit crab. Strictly seen during the summer season and widely seen during the lowtide region at outer channel area.

Distribution in India: Odisha, Mahanadi Estuary; Chilika Lake, Gopalpur; Chhatrapur; Rushikulya estuary; Hukitola; Chandipur; **West Bengal, Andhra Pradesh, Gujarat** : Gulf of Kachchh,, **Kerala**,, **Tamil Nadu** : Ennore Backwaters, Gulf of Mannar; Krusadai; Shingle Island, Kerla, Kochi; Kozhikode, **Karnataka**: Karwar, **Maharashtra**: Mumbai, **Andaman and Nicobar Islands.**

Remark: This is the first time ever to record this species from Chilika lake. It is a true marine species occurrence. This species is uncommon to the Odisha coast as well as to Chilika lake. It is first time to record from the lagoon premises after the opening of the artificial lagoon inlet. Initially it is believed that *B. Spirata* might be washed in to the lake due to the tidal incursion of wave action . But, frequent occurrence of live shells confirmed that this species already accommodated in the lake ecosystem.

Babylonia zeylanica

Phylum : Mollusca

Class: Gastropoda

Order : Neogastropoda

Family : Babyloniidae

Genus and species : *Babylonia zeylanica* Bruguiere, 1789

Babylonia zeylanica

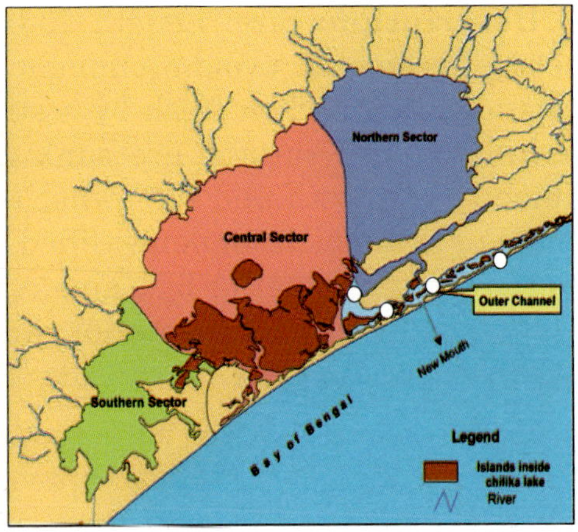

Distribution in Chilika lagoon

Description Size: Up to 70 mm in height, Shell large, fusiform, less solid and with less inflated whorls, body whorl narrower than in *B. spirata,* sutures not canaliculated, spire high ending in dark purple apex, Aperture large, outer lip sharp and smooth, but not flexed at top, columella smooth, with heavy broad callus posteriorly but narrow anteriorly, a strong parietal ridge almost close to the outer lip, umbilicus broadly open with a row of teeth on the outer margin, fasciole with a ridge on the inner edge, anterior canal broad and deep, posterior canal not distinct. Surface smooth.

Body color: White with large brown blotches.

Distribution in Chilika: In Chilika lake this marine gastropod is recorded from outer channel area only. Some dead shells are occupied by the hermit crab. Strictly seen during the summer season and widely seen during the low tide region at outer channel area.

Distribution in India: Odisha, Mahanadi Estuary; Chilika Lake, Gopalpur; Chhatrapur; Rushikulya estuary; Hukitola; Chandipur; **West Bengal, Andhra Pradesh, Gujarat** : Gulf of Kachchh,, **Kerala**,, **Tamil Nadu** : Ennore Backwaters, Gulf of Mannar; Krusadai; Shingle Island, Kerla, Kochi; Kozhikode, **Karnataka**: Karwar, **Maharashtra**: Mumbai, **Andaman and Nicobar Islands.**

Remark: This is the first time ever to record this species from Chilika lake. It is a true marine species occurrence. This species is uncommon to the Odisha coast as well as to Chilika lake. It is first time to record from the lagoon premises after the opening of the artificial lagoon inlet. But, frequent occurrence of live shells confirmed that this species already accommodated in the lake ecosystem.

Bellamya bengalensis

Phylum : Mollusca

Class: Gastropoda

Order : Architaenioglossa

Family : Viviparidae

Genus and species : *Bellamya bengalensis* Lamarck, 1822

Bellamya bengalensis

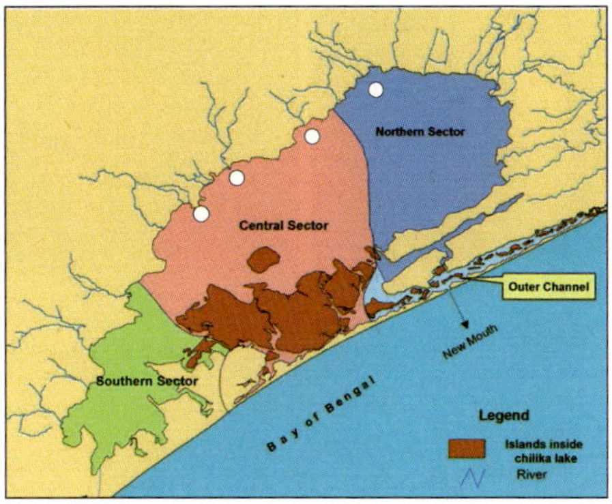

Distribution in Chilika lagoon

Description: Shell elongate, with dark spiral bands; sutures not greatly impressed; aperture sub-circular; Body whorl perfectly rounded with conical spire .

Distribution in Chilika : Found in the fresh water dominated region of Chilika such as Kalupadaghat, Bhusandapur, Tinimuhani, Soran etc.

Distribution in India: Commonly seen throughout India, in the fresh water ponds, lakes, river mouths etc.

Remark: Used as food material by the local people, as a diet for the aquaculture practice. During the monsoon season their abundance and biomass shows an increased value. But sharp decline was noticed as the salinity of the study area increases gradually. During summer season the when the average salinity of the lake area is high at its peak (> 10 ppt) then this species showed high rate of mortality.

Bullia vittata

Phylum : Mollusca

Class: Gastropoda

Order : Neogastropoda

Family : Nassariidae

Genus and species: *Bullia vittata* Linnaeus, 1767

Bullia vittata

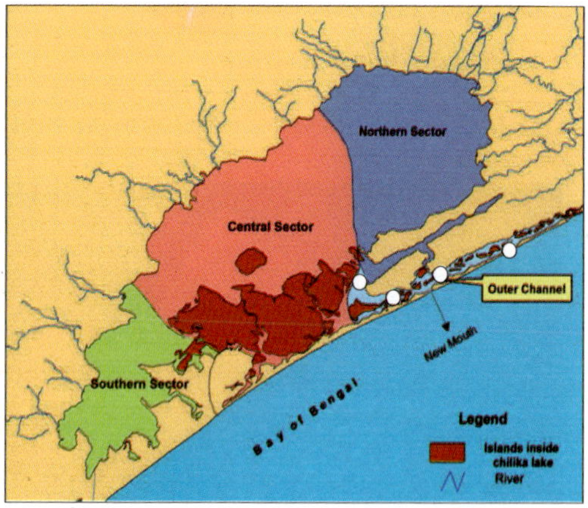

Distribution in Chilika lagoon

Description Size: Length 6.75-53.69 mm; diameter 2.84-19.74 mm; height of spire 4.98-36.43 mm.Shell turreted, spire elongate, longer than body whorl; beaded spiral cords below suture. It has a massive foot, with the help of which it quickly burrows into the sand. Aperture wide; outer lip thin; columella smooth; anterior siphonal notch very wide. Sculptured with spiral grooves separated by wide interspaces, three rows of beaded spiral cords just below the sutures.

Distribution in Chilika : Found in the marine dominated region of Chilika such as Satapada, Rambharatia, Bhavakundaleswar, Arakhkuda.

Distribution in India : Odisha, Mahanadi Estuary; Chilika Lake, Gopalpur; Chhatrapur; Rushikulya estuary; Hukitola; Chandipur; **West Bengal**: Jharkhali; North 24 Parganas; Sundarbans; Champta. **Andhra Pradesh,** Chittivalasa; **Gujarat** : Gulf of Kachchh, **Karnataka**, **Kerala**, Maharashtra, **Tamil Nadu** : Ennore Backwaters, Gulf of Mannar; Krusadai; Shingle Island, Kerla, Kochi; Kozhikode, Karnataka: Karwar, **Maharashtra**: Mumbai, Lakshadweep, Andaman and Nicobar Islands.

Abroad: Bangladesh, Malaysia.

Remark: This species is estuarine and intertidal can live within the salinity range of 20 ppt to 35 ppt embedded in sandy substratum. In many parts of the world, it is used as food item for human consumption and also used as bait to catch fish and shell fish. This species is also used for the extraction of lime.

They crawl on the substratum in order to feed on settled planktonic materials. This is also used as an indicator of heavy metal pollution, oil pollution in the many parts of the world.

Bursa spinosa

Phylum : Mollusca

Class: Gastropoda

Order : Hygrophila

Family : Bursidae

Genus and species : *Bursa spinosa* Lamarck, 1843

Bursa spinosa

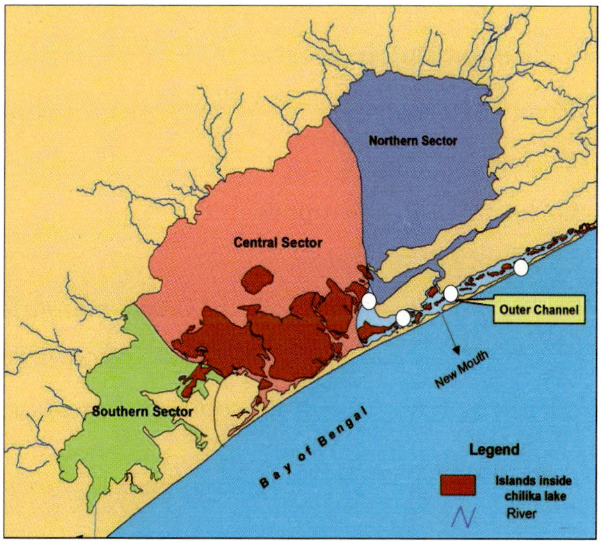

Distribution in Chilika lagoon

Description: Length: 18.15-70.00 mm,Diameter: 14.25-38.45 mm,Max height of the aperture: 9.50-28.10 mm. Shell dorsoventrally depressed, slightly angulated whorls. Shell contains strong spines; single spine adjacent to the posterior canal directed backwards. One or two regular, spiral rows of widely spaced, short, pointed, spine like tubercles. The edge of the outer lip is strong, irregular, appeared with transverse teeth.

Body Color: The shell is uniformly smoky brown.

Distribution in Chilika : Found in the marine dominated region of Chilika such as Satapada, Rambharatia, Bhavakundaleswar, Arakhkuda .

Distribution in India : Odisha, Mahanadi Estuary; Chilika Lake, Gopalpur; Chhatrapur; Rushikulya estuary; Hukitola; Chandipur; **West Bengal**: Jharkhali; North 24 Parganas; Sundarbans; Champta. **Andhra Pradesh,** Chittivalasa; **Gujarat** : Gulf of Kachchh, **Karnataka**, **Kerala**, Maharashtra, **Tamil Nadu** : Ennore Backwaters, Gulf of Mannar; Krusadai; Shingle Island, Kerla, Kochi; Kozhikode, Karnataka: Karwar, **Maharashtra**: Mumbai, **Lakshadweep, Andaman and Nicobar Islands.**

Abroad: Bangladesh, Malaysia.

Remark: This species is estuarine and intertidal can live within the salinity range of 20 ppt to 35 ppt embedded in sandy substratum. In many parts it used as food item for human consumption and also used as bait to catch fish and shell fish.

Cerithideopsilla cingulata

Phylum : Mollusca

Class: Gastropoda

Order : Caenogastropoda

Family : Potamididae

Genus and species : *Cerithideopsilla cingulata* Gmelin 1791

Cerithideopsis costata da Costa, 1778

Telescopium telescopium Linnaeus, 1758

Cerithideopsilla cingulata

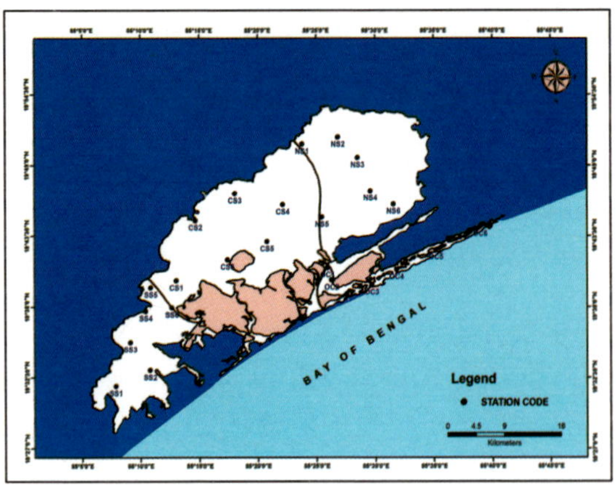

Distribution in Chilika lagoon

Description Size: 12.32-23.41 mm, Shell moderately large, narrowly elongate and thick; 13-15 flat sided whorls separated by shallow sutures; aperture oval; columella almost straight; outer lip thick and expanded broadly; anterior siphonal canal distinct and short with a varice-like bulge and blunt rib above; sculptured with spiral ridges crossed by strong axial ribs forming rows of regular granular nodules.

Body color: Color: Dark brown, often with a whitish band above the suture, nodules dirty white, interstices brown, interior of aperture white, lined with brown.

Distribution in India: This species is commonly recorded from the east coast and west coast of India. However some of the conspicuous ecosystems are those of **Orissa**: Fals point; Chandipur; Dhamra estuary; Paradwip; Mahanadi Estuary; Konark; Puri; Chilika lake; Ruashikulya estuary; Chhatrapur; Gopalpur, **West Bengal**: Diamond Harbour; Sundarbans,; Digha Coast; Ganga Sagar; Port Canning; Medinipur; North 24 Parganas; South 24 Parganas; Sankarpur; Old Digha, **Gujarat**, **Kerala**,, Maharashtra: Malvan; Ratnagiri; Mumbai; Mahim Creek; Alibag; Nariman Point; Kalbadevi Estuary; Bhatia Creek; Tarkarli Creek, **Goa**: Siridao beach; Siolim; Ribandar; Banastrim; Ribandar; Mandovi Estuary, **Karnataka**: North Kanara; Karwar; Kodibag; Ankola; Tadri, **Pondicherry**: Pondicherry mangrove; **Tamil Nadu**: Parangipettai Coast; Gulf of Mannar; Arukkattuthurai; Kodikkarai; Mallipattinam; Manalmelkudi; Adirampattinam; Adyar Estuary; Vellar Estuary; Cooum Estuary; Pulicat Lake; Pamban Kerala; Porto Novo, Tellicherry; Karangadu; Athankarai; Ennore; Thondi; Karangadu , **Andaman and Nicobar Islands**.

Remark: In Chilika found in sandy intertidal bed, mixed sediment, sea grass beds, sea weeds beds.

Chicoreus torrefactus

Phylum : Mollusca

Class: Gastropoda

Order : Neogastropoda

Family : Muricidae

Genus and species : *Chicoreus torrefactus* (Lamarck, 1816)

Chicoreus torrefactus

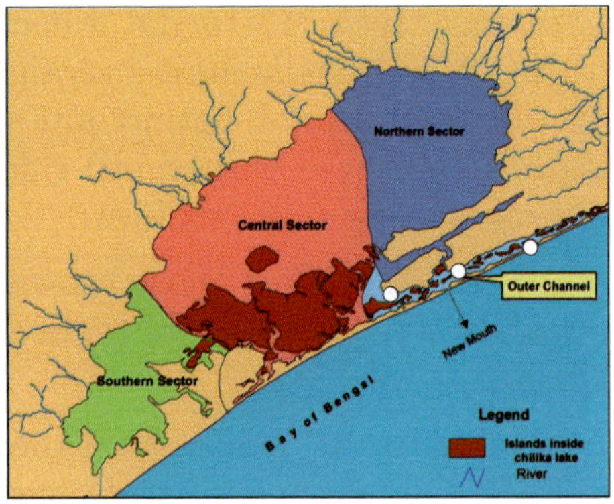

Distribution in Chilika lagoon

Description Size: Length 60.26-91.10 mm; diameter 31.25-42.75 mm; length of aperture 15.14-28.05 mm; length of spire 27.26-49.85 mm; length of the canal 19.28-23.65 mm. Shell fusoid, large; spire high consisting of nine post nuclear whorls. Protoconch consists of 2½ whorls, rounded, depressed and smooth; body whorl large, elongate; suture not deep; anal sulcus narrow, deep. Aperture is circularly ovate; outer lip thick and dentate.

Body colour: Light to dark brown with varices and spiral cords chestnut; aperture white with light pink tinge.

Distribution in Chilika : It is recorded from the outer channel area during March to June.

Distribution in India: This species is rare along north east coast of India although common in the south east coast. Previously this was recorded from Gopalpur; Odisha, Chnneai, **Tamil Nadu**:, **Andaman and Nicobar Islands** and **Lakshadweep.**

Remark: Rare in occurrence, only seen from the outer channel are as live specimen. This species have a great economic value since widely demanded for the decorative purposes. Dead shells are collected for the lime production and also for the auspicious uses.

Chicoreus virginus

Phylum : Mollusca
Class: Gastropoda
Order : Neogastropoda
Family : Muricidae
Genus and species : *Chicoreus virginus*

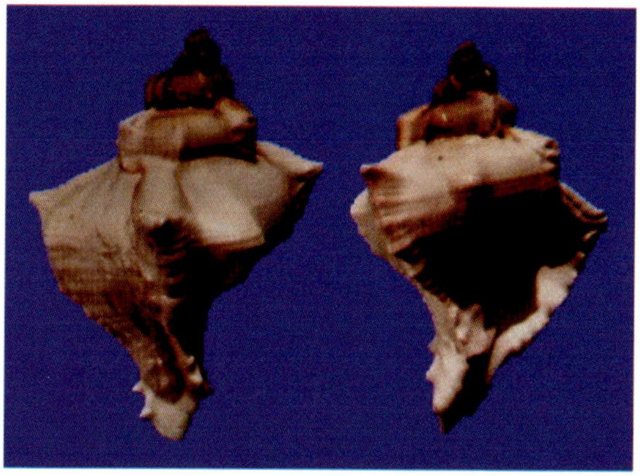

Chicoreus virginus

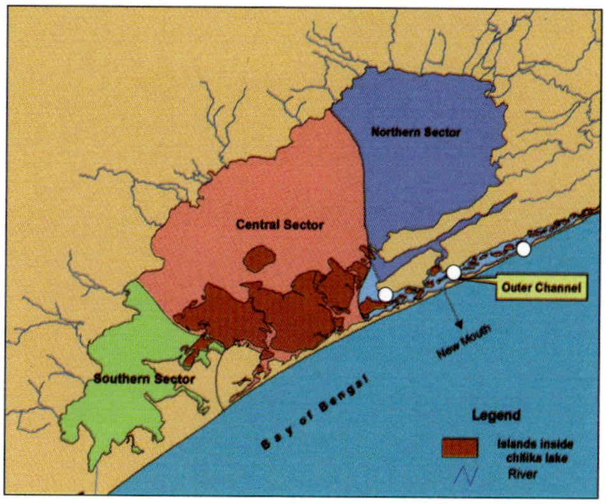

Distribution in Chilika lagoon

Description: The anterior canal is moderately long and is partially closed by a broad, horizontal fold from the basal prolongation of the body whorl. The outer surface is pale brown. The shell is not only much thicker, heavier and more massive, but also relatively broad in proportion to its total length. The surface is traversed by close-set, fine, unequal spiral ridges, most of which are minutely granulated. The wide interspaces between the varices bear tubercles. The varices are much stouter and more raised and their folded processes are few, poorly developed and widely spaced. The columella bears a smooth, polished callus layer.

Body Color: The columella and the outer lip are edged with pink, while the rest of the interior of the aperture is whitish.

Distribution in Chilika: In Chilika lake this marine gastropod is recorded from outer channel area only. Some dead shells are occupied by the hermit crab. Strictly seen during the summer seaon and widely seen during the lowtide region at outer channel area.

Distribution in India: Odisha, Mahanadi Estuary; Chilika Lake, Gopalpur; Chhatrapur; Rushikulya estuary; Hukitola; Chandipur; **West Bengal, Andhra Pradesh, Gujarat** : Gulf of Kachchh,, **Kerala**,, **Tamil Nadu** : Ennore Backwaters, Gulf of Mannar; Krusadai; Shingle Island, Kerla, Kochi; Kozhikode, **Karnataka**: Karwar, **Maharashtra**: Mumbai, **Andaman and Nicobar Islands.**

Remark: This a new record to Chilika lake observed after the opening of new lagoonal inlet.

Fusinus colus

Phylum:Mollusca

Class:Gastropoda

Order:Neogastropoda

Family:Fasciolariidae

Genus-Species - *Fusinus colus* Linnaeus, 1758

Fusinus colus

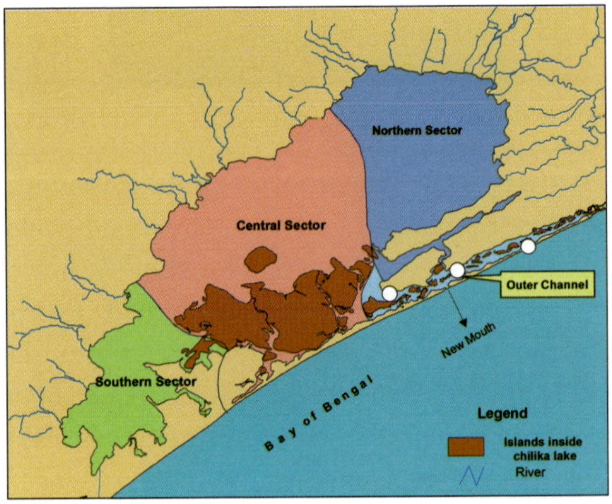

Distribution in Chilika lagoon

Description Size: Length 78.36-104.4 mm; diameter 18.50-23.48 mm. Elongate, spire high, siphonal canal very long and straight, early whorls slightly angulate with a central row of nodules, later whorls rounded, the first 3-4 whorls with sculptured throughout with rough spiral cords columellar callus smooth or so thin that the cords beneath are evident.

Body colour: White, tip of canal brown, with brown patches between the nodules on the spire.

Distribution in Chilika : Found in the marine dominated region of Chilika such as Rambharatia, Bhavakundaleswar, Arakhkuda.

Distribution in India : Odisha, Mahanadi Estuary; Chilika Lake, Gopalpur; Chhatrapur; Rushikulya estuary; Andhra Pradesh, Karnataka, Kerala, Maharashtra, Tamil, Lakshadweep, Andaman and Nicobar Islands.

Abroad: Srilanka

Remark: This species is estuarine and intertidal can live within the salinity range of 20 ppt to 35 ppt embedded in sandy substratum. In many parts it used as food item for human consumption and also used as bait to catch fish and shell fish. They crawl on the substratum in order to feed on settled planktonic materials. Dead shells are used in the decorative purposes and for the preparation of lime. In Chilika Lake this is frequently observed in the high salinity period i.e. during the summer season at the outer channel area. When dead its muscle feed on by the shore birds and other carnivorous gastropods even by the small fishes. Thus play a critical role in the coastal food chain.

Harpa davidis

Phylum : Mollusca

Class: Gastropoda

Order : Neogastropoda

Family : Harpidae

Genus and species : *Harpa davidis* Roding, 1798

Harpa davidis

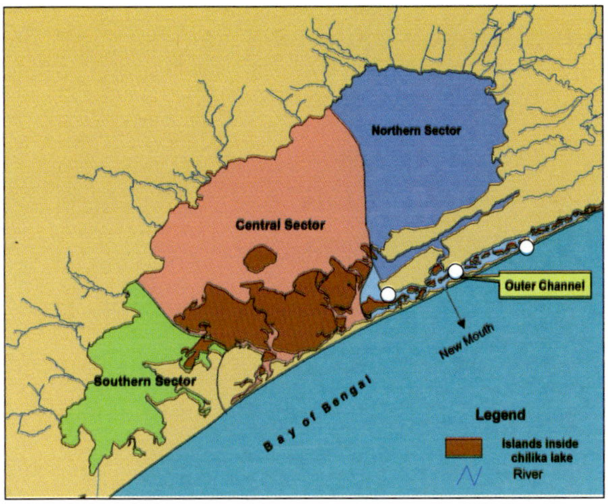

Distribution in Chilika lagoon

Description Size: Length 42.56-63.3 mm; diameter 27.92-44.25 mm.Shell moderate size, broadly ovate with rather short, broadly conical spire. Protoconch elevated-mamillate, pale pink, of 3 ¼ to 3 ¾ convex whorls. Early postnuclear whorls reticulated by axial riblets and spiral cords, moderately angulate of antepenultimate whorl covered by extensions of earlier parietal calluses.

Body Color: Ground color grayish pink arrangement, red brown or pale pink spiral bands pattern.

Distribution in Chilika : Found in the marine dominated region of Chilika such as Satapada, Rambharatia, Bhavakundaleswar, Arakhkuda.

Distribution in India: Odisha, Mahanadi Estuary; Gopalpur; Chandipur; **Tamil Nadu** : Tharangambadi; Pamban; Chennai; Rameswaram, **Andaman and Nicobar Islands**.

Abroad: Bangladesh , Malaysia.

Remark: This species is estuarine and intertidal can live within the salinity range of 20 ppt to 35 ppt embedded in sandy substratum. Iit used as food item for human consumption and also used as bait to catch fish and shell fish. They crawl on the substratum in order to feed on settled planktonic materials. Dead shells are used in the decorative purposes and for the preparation of lime. In Chilika lake only dead shells re found till date. It is belived that it might have been carried out by the ghost crab or through high wave action from Bay of Bengal. This species is not reported before the opening of the new lagoon inlet opened by artificial means during the year 2000.

Indoplanorbis exustus

Phylum : Mollusca

Class: Gastropoda

Order : Hygrophila

Family : Planorbidae

Genus and species : *Indoplanorbis exustus* Deshayes, 1834

Indoplanorbis exustus

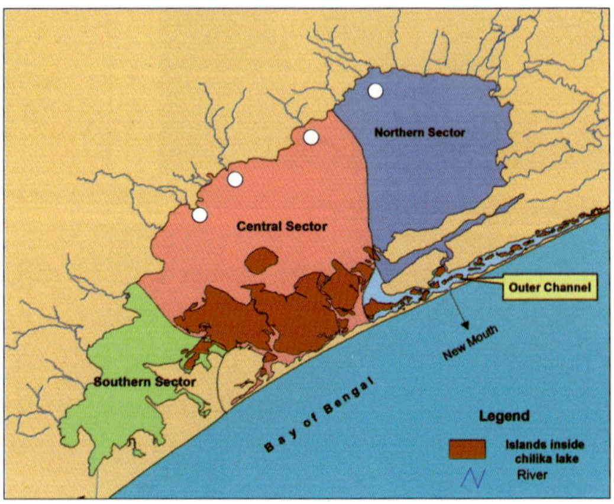

Distribution in Chilika lagoon

Description: The shell of this species, like all planorbids is sinistral in coiling, but is carried upside down and thus appears to be dextral. The shell of *Indoplanorbis exustus* is discoid with rapidly increasing whorls. Each whorl is higher than it is wide. The width of the shell is 5-25 mm. The height of the shell is 4.5-13 mm.

Distribution in Chilika : Fresh water dominated regions of northern sector and central sectors, river mouth regions.

Distribution in India: Bihar, Haryana, Jharkand, Madhya Pradesh, Rajasthan, Uttaranchal, Uttar Pradesh, West Bengal.

Abroad: Iran, Saudi Arabia, Oman and Socotra Island in Yemen, China, Malay Peninsula and Archipelago, Myanmar, Pakistan, Persia, Sumatra, Celebes, Java, Malaya, Thailand, Indonesia, Malaysia, Vietnam, Bangladesh, India, Sri Lanka and Nepal.

Remark: It is a common freshwater bivalve species often located in rivers, streams and lakes, prefers to live in fine gravel, sand or mud substratum; mostly confined to lowland rivers up to around 250 m used as food in many parts of Nepal, India and Bangladesh. In Chilika it embeds in the river mouth bed in the Kalupadaghat region, in Sorana, Tinimuhani, Bhusandapur region etc. It provides a great socio-economic support to the local fishermen since they used it as food as the food particles for prawn culture.

Indothais lacera

Phylum : Mollusca

Class: Gastropoda

Order : Neogastropoda

Family : Muricidae

Genus and species : *Indothais lacera* Born, 1778

Indothais lacera

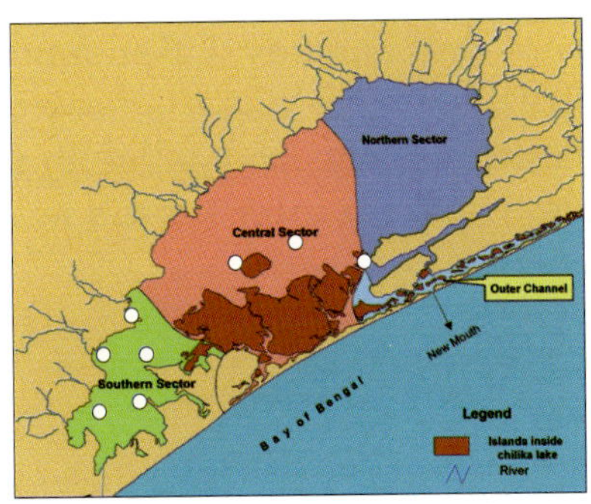

Distribution in Chilika lagoon

Description Size: Length 15.25-25.52 mm; diameter 17.80-35.55 mm; length of aperture 15.24-28.40 mm; length of spire 10.95-26.32 mm.Whorls sharply angulate, carinate; body whorl with two rows of spinose tubercles, surface with spiral ribs and groves, not scabrous; aperture lirate inside on outer lip; columella smooth. Anal sulcus deep inverted 'U' shaped, margined by parietal ridge. Sculpture consists of transverse striations flat, mildly scabrous, two rows of tubercles on upper part of the body whorl, sometimes much prominent like spines sometimes single row of tubercles present.

Body Color: Ashy brown, aperture light brown, margin white.

Distribution in Chilika : In Chilika lake this species is recorded from Plaur outfall point, Pathara, Samal nashi, Badakuda, Sanakuda, Talatalaanashi. Budhibaranashi, Krusnaprasad gargh, Nalabana Is. Magarmukh, new inlet region, Old inlet region (Arakhkuda).

Distribution in India: This species is also recorded from **Orissa**: Subarnarekha Mouth; Talasari, Kirtania; Chandipur; Dhamra estuary; Paradwip, Mahanadi Estuary, Konark; Chilika lake, Ruashikulya estuary, Chhatrapur, **West Bengal**: Diamond Harbour; Sundarbans,; Digha Coast; Ganga Sagar; Port Canning; Medinipur; North 24 Parganas; South 24 Parganas; Sankarpur; Old Digha; **Andaman and Nicobar Islands**, **Gujarat**, **Kerala**,, **Maharashtra**: Malvan; Ratnagiri; Mumbai; Mahim Creek; Alibag; Nariman Point; Kalbadevi Estuary; Bhatia Creek; Tarkarli Creek, Goa: Siridao beach, Siolim, Ribandar, **Karnataka**: North Kanara; Karwar; Kodibag; Ankola; Tadri, **Pondicherry**: Pondicherry mangrove; **Tamil Nadu**: Parangipettai Coast; Gulf of Mannar; Arukkattuthurai; Kodikkarai; Mallipattinam; Manalmelkudi; Adirampattinam; Adyar Estuary; Vellar Estuary; Cooum Estuary; Pulicat Lake; Pamban Kerala; Porto Novo, Tellicherry.

Abroad: Circum-tropical i.e. from Africa to Japan.

Remark: This species is typically constituent of the entire southern sector of Rambha bay. Often found attached in the sea grass or crawling of the substratum of the lake. Its flesh is eaten by the birds. Flesh is also used for the preparation of the food pellet for the aquaculture firms. Not edible by the locale people. During winter season i.e. from October to February higher abundance is noticed.

Murex tribulus

Phylum : Mollusca

Class: Gastropoda

Order : Neogastropoda

Family : Muricidae

Genus and species : *Murex tribulus* Linnaeus, 1758

Murex tribulus

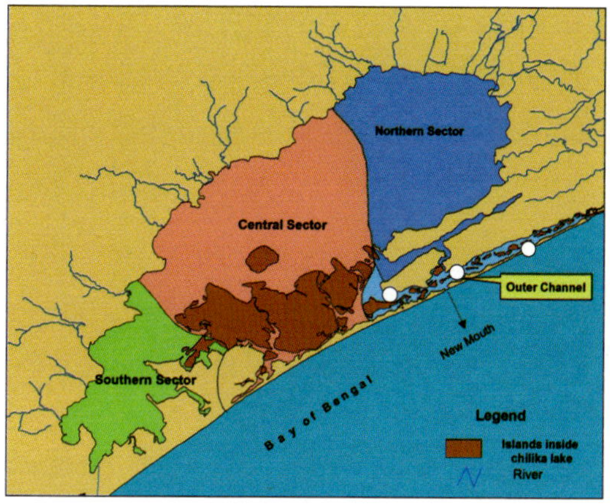

Distribution in Chilika lagoon

Description Size: 10 cm; length: 30.80-41.14 mm; diameter: 12.20-16.15 mm; height of the aperture: 12.05-8.65 mm Shell club shaped; Spire acute with 6-7 whorls; Body whorl globose; Varices with long spire; Aperture large, outer lip erect with limited crenulations; Canal elongate and narrow.

Distribution in Chilika : In Chilika lake this marine gastropod is recorded from outer channel area only. Some dead shells are occupied by the hermit crab. Strictly seen during the summer season and widely seen during the lowtide region at outer channel area.

Distribution in India:. It is also recorded from **Orissa**: Subarnarekha Mouth; Talasari, Chandipur, Dhamra estuary; Paradwip, Mahanadi Estuary, Konark; Chilika lake, Ruashikulya estuary, Chhatrapur, **West Bengal**: Diamond Harbour; Sundarbans,; Digha Coast; Ganga Sagar, **Andaman and Nicobar Islands**, **Gujarat**, **Kerala**,, **Maharashtra**, **Tamil Nadu**: Parangipettai Coast; Gulf of Mannar; Arukkattuthurai; Kodikkarai; Mallipattinam; Manalmelkudi; Adirampattinam; Adyar Estuary; Vellar Estuary; Cooum Estuary; Pulicat Lake; Pamban; Porto Novo, Tellicherry.

Abroad: Red sea to Philipines and Japan.

Remark: This species is common representative all along the east coast and west coast of India. Dead shells are used as the decorative purposes as well as for the lime formation.

Nassarius stolatus

Phylum : Mollusca

Class: Gastropoda

Order : Neogastropoda

Family : Nassariidae

Genus and species : *Nassarius stolatus* Gmelin, 1791
N. foveolatus Dunker, 1847
N. subconstrictus Sowerby, 1899
N. jacksonianus Quoy & Gaimard, 1833
N. orissaensis Preston, 1914

Nassarius stolatus

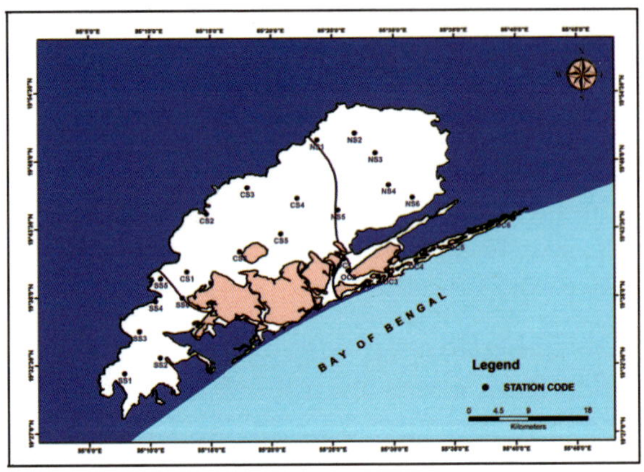

Distribution in Chilika lagoon

Description: Measurement: L = 7-22, D = 4-14; Shell small, Yellowish white in colour with three broad chest nut spiral bands, ovately conical in shape; Inflated body whorl; pointed spire, few spire cords below suture.

Distribution in Chilika : In Chilika lake this species is cosmopolitan in nature. Found all most all the regions of the lake ecosystem starting from sea grass bed to macrophyte infested region. This species is found everywhere with in the lake premises. It is one among the dominant species.

Distribution in India: India : **Orissa** : Subarnarekha estuary, Chilika lagoon, Mahanadi Estuary (Hukitola); **Andhra Pradesh**: Kakinada Bay, Godavari Estuary, **Maharashtra** :Bombay; **Tamil Nadu** : Tranquebar, Pamban, Chennai and **West Bengal**: Kolkata, Sagar Island, Digha, Hugli Matla Estuary.

Abroad: Gulf of Aden; Persias gulf, Red Sea to Philippines and to Japan.

Remark: It has a great ecological importance value since it is being grazed by the crabs, prawn and even by the migratory birds thus play critical role in the lagoon food chain. It is found throughout the year in the lake substratum.

Natica gualteriana

Phylum : Mollusca

Class: Gastropoda

Order : Littorinimorpha

Family : Naticidae

Genus and species : *Natica gualteriana* Récluz, 1844
N. vitellus Linnaeus, 1758
N. tigrina Röding, 1798
Eunaticina coarctata Reeve 1864

Natica gualteriana

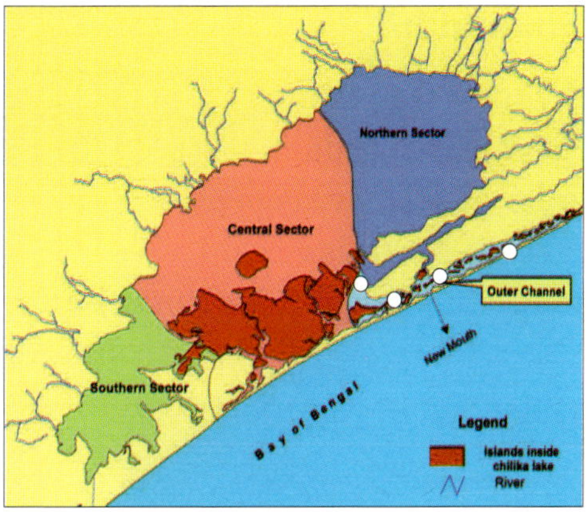

Distribution in lake

Description Size: Height 7.05 - 28.20 mm, maximum diameter 10.20-43.30 mm, minimum diameter 8.45 - 34.20 mm. Shell is thick, globose, excavately umbilicated with an inflated body whorl; spire reduced; whorls rounded; surface obliquely plicately striated; columella smooth, white with polished callus extended posteriorly and fused with the body whorl leaving anteriorly the umbilical opening; aperture lunar ovate.

Body Colour: Fulvous-brown, pale band in the middle and whitish in the base and the sutures.

Distribution in Chilika : Found in the marine dominated region of Chilika such as Satapada, Rambharatia, Bhavakundaleswar, Arakhkuda .

Distribution in India : Odisha, Mahanadi Estuary; Chilika Lake, Gopalpur; Chhatrapur; Rushikulya estuary; Hukitola; Chandipur; **West Bengal**: Jharkhali; North 24 Parganas; Sundarbans; Champta. **Andhra Pradesh,** Chittivalasa; **Gujarat** : Gulf of Kachchh, **Karnataka**, **Kerala**, Maharashtra, **Tamil Nadu** : Ennore Backwaters, Gulf of Mannar; Krusadai; Shingle Island, Kerla, Kochi; Kozhikode, Karnataka: Karwar, **Maharashtra**: Mumbai, **Lakshadweep, Andaman and Nicobar Islands.**

Abroad: Bangladesh , Malaysia

Remark: This species is estuarine and intertidal can live within the salinity range of 20 ppt to 35 ppt embedded in sandy substratum. In many parts it used as food item for human consumption and also used as bait to catch fish and shell fish. They crawl on the substratum in order to feed on settled planktonic materials. Dead shells are used in the decorative purposes. In Chilika lake this is frequently observed in the high salinity period i.e. during the summer season at the outer channel area.

Neritina smithii

Phylum : Mollusca

Class: Gastropoda

Order : Cycloneritimorpha

Family : Neritidae

Genus and species : *Neritina smithii* Wood, 1828
Nerita polita Reeve, 1855

Neritina smithii

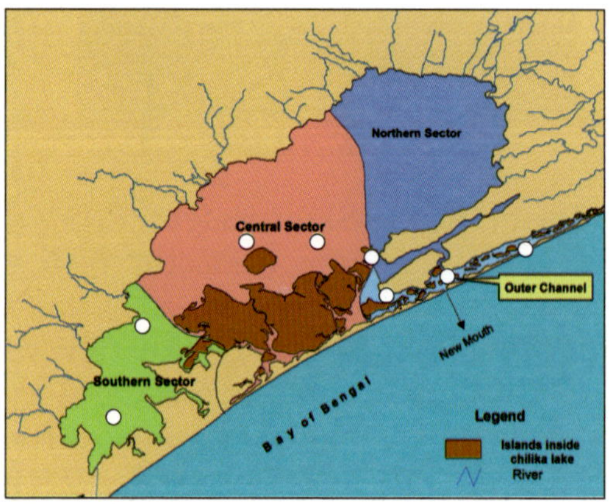

Distribution in Chilika lagoon

Description Size: 3 cm wide. Subglobose, diameter greater than height, spire flat, surface shiny but sculptured with growth lines, columellar deck smooth but with several weak, rounded, marginal teeth at the centre, operculum smooth except for a finely radialy striate zone near the margin. **Body Color**: Cream, brown, pink or green, with irregular black or brown markings, aperture orange, columellar deck red or dark orange, operculum black, grey.

Distribution in Chilika: Found in the marine dominated region of Chilika such as Satapada, Rambharatia, Bhavakundaleswar, Arakhkuda .

Distribution in India : Odisha, Mahanadi Estuary; Chilika Lake, Gopalpur; Chhatrapur; Rushikulya estuary; Hukitola; Chandipur; **West Bengal**: Jharkhali; North 24 Parganas; Sundarbans; Champta. **Andhra Pradesh,** Chittivalasa; **Gujarat** : Gulf of Kachchh, **Karnataka**, **Kerala**, Maharashtra, **Tamil Nadu** : Ennore Backwaters, Gulf of Mannar; Krusadai; Shingle Island, Kerla, Kochi; Kozhikode, Karnataka: Karwar, **Maharashtra**: Mumbai, **Lakshadweep, Andaman** and **Nicobar Islands.**

Abroad: Bangladesh , Malaysia.

Remark: This species is estuarine and intertidal can live within the salinity range of 20 ppt to 35 ppt embedded in sandy substratum. In many parts it used as food item for human consumption and also used as bait to catch fish and shell fish. They crawl on the substratum in order to feed on settled planktonic materials. Dead shells are used in the decorative purposes. In Chilika lake this is frequently observed in the high salinity period i.e. during the summer season at the outer channel area.

Oliva oliva

Phylum : Mollusca

Class: Gastropoda

Order : Neogastropoda

Family : Olividae

Genus and species : *Oliva oliva* Linnaeus, 1758

Oliva oliva

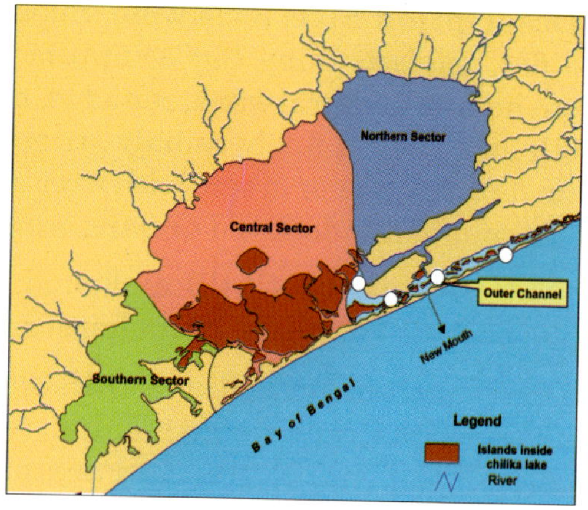

Distribution in Chilika lagoon

Description: Size: 4 cm, Sides somewhat convex, spire moderately low, fasciole poorly defined, columella thickly calloused, callus extending to slightly more than half way along the aperture, moderately lirate, with three strong, angulate spiral folds anteriorly.

Distribution in Chilika : In Chilika lake this species is recorded from Magarmukh, new inlet region, Old inlet region (Arakhkuda).

Body Colour: Cream to brown with brown or purplish spots, zigzag lines and sometimes spiral bands, fasciole sometimes with indistinct crescent-shaped brown lines, interior pale brown or blue, columella white.

Distribution in India: Distribution in India: This species is commonly recorded from the east coast and west coast of India. However some of the conspicuous ecosystems are those of **Orissa**: False point; Chandipur; Dhamra estuary; Paradwip; Mahanadi Estuary; Konark; Puri; Chilika lake; Ruashikulya estuary; Chhatrapur; Gopalpur, **West Bengal**: Diamond Harbour; Sundarbans,; Digha Coast; Ganga Sagar; Port Canning; Medinipur; North 24 Parganas; South 24 Parganas; Sankarpur; Old Digha, **Gujarat**, **Kerala**,, **Maharashtra**: Malvan; Ratnagiri; Mumbai; Mahim Creek; Alibag; Nariman Point; Kalbadevi Estuary; Bhatia Creek; Tarkarli Creek, **Goa**: Siridao beach; Siolim; Ribandar; Banastrim; Ribandar; Mandovi Estuary, **Karnataka**: North Kanara; Karwar; Kodibag; Ankola; Tadri, **Pondicherry**: Pondicherry mangrove; **Tamil Nadu**: Parangipettai Coast; Gulf of Mannar; Arukkattuthurai; Kodikkarai; Mallipattinam; Manalmelkudi; Adirampattinam; Adyar Estuary; Vellar Estuary; Cooum Estuary; Pulicat Lake; Pamban Kerala; Porto Novo, Tellicherry; Karangadu; Athankarai; Ennore; Thondi; Karangadu , **Andaman and Nicobar Islands**.

Abroad: Circum-tropical i.e. from Africa to Japan.

Remark: This species is typically constituent of the entire southern sector of Rambha bay. Often found attached in the sea grass or crawling of the substratum of the lake. Its flesh is eaten by the birds and the dead shell is used for the production of lime. Flesh is also used for the preparation of the food pellet for the aquaculture firms. Not edible by the locale people. During winter season i.e. from October to February higher abundance is noticed.

Olivancillaria gibbosa

Phylum : Mollusca

Class: Gastropoda

Order : Neogastropoda

Family : Olividae

Genus and species : *Olivancillaria gibbosa* Born, 1778

Olivancillaria gibbosa

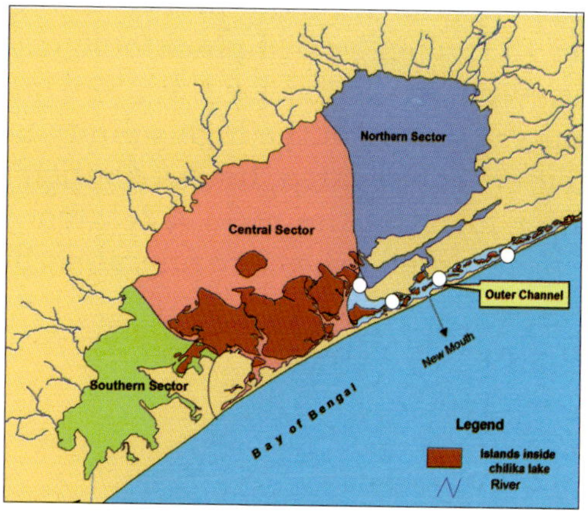

Distribution in Chilika lagoon

Description: The shell is stout, thick with a smooth and highly polished surface and an elongately ovoid, somewhat inflated body whorl, the inflation being more strongly marked towards the upper part. The spire is rather short with a pointed apex. The columella bears a dense, polished white callus thickening which is particularly well marked posterior while anteriorly it bears numerous, oblique close-set ridges. The aperture is narrow and elongated, with a small slit like posterior canal and a broadly opened anterior canal in the form of a semilunar notch. **Body Colour**: The upper part of the body whorl ashy blue with pale brownish marking; lower part which is generally sharply demarcated from the upper by an oblique spiral line, yellowish white dark brown patches.

Distribution in Chilika lake : Old mouth (Arakhkuda), New mouth (Shipakuda), Satapada, Rambharatia Is.

Distribution in India : Odisha, Mahanadi Estuary; Chilika Lake, Gopalpur; Chhatrapur; Rushikulya estuary; Subarnarekha estuary; **Andhra Pradesh, Gujarat** :; Mahuva; Veraval; Okha Reef; Gujarat, Hanumandandi Reef; Gulf of Kachchh,, **Kerala**, Kochi, Sakthikulangara; Neendakara, **Tamil Nadu** : Palk Bay, Gulf of Mannar; Mandapam; Parangipettai Coast, **Karnataka**: Karwar, **Maharashtra**: Mumbai,**Goa**: Candolim beach **Andaman and Nicobar Islands.**

Abroad: Circum-tropical.

Remark: This species is estuarine and intertidal can live within the salinity range of 20 ppt to 35 ppt embedded in sandy substratum. Usually seen in the outer channel area only during the low tide periods of summer seasons. Its availability in the lake proper is very less. It crawl on the substratum in order to feed on settled planktonic materials. Dead shells are used in the decorative purposes and for the preparation of lime. In Chilika lake this is frequently observed in the high salinity period i.e. during the summer season at the outer channel area. Its distribution is restricted to Satapada region only and never found in the main lake area . It's body muscle is important food source fishes, shell fishes and migratory birds.

Phalium areola

Phylum : Mollusca

Class: Gastropoda

Order : Littorinimorpha

Family : Cassidae

Genus and species : *Phalium areola* Linnaeus, 1758

Phalium areola

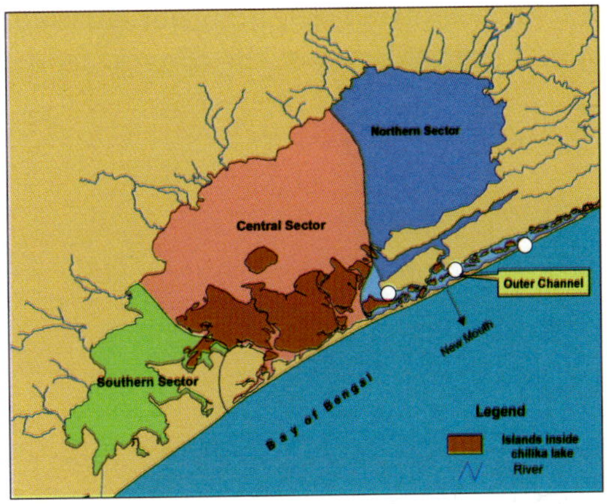

Distribution in Chilika lagoon

Description Size: 90 mm in length,Outer lip thick and recurved, broader especially towards the base, with 22 widely separated teeth, lower half of columella and parietal shield with irregular coarse, spiral ridges anterior canal deep and slightly deflected to the left. Spire and shoulder of penultimate whorl cancellate, without any nodes, body whorl smooth and glossy with a well-developed varix. **Body Colour**: Color: White with 5 rows of rhomboidal brown spots and outer edge of shield with pink-brown tinge.

Distribution in Chilika lake : Old mouth (Arakhkuda), New mouth (Shipakuda), Rambharatia Is.

Distribution in India : Odisha, Mahanadi Estuary; Chilika Lake, Gopalpur; Chhatrapur; Rushikulya estuary, **Andhra Pradesh, Gujarat** : Gulf of Kachchh,, **Kerala**, **Tamil Nadu** : Ennore Backwaters, Gulf of Mannar; **Kerla**, Kochi; **Karnataka**: Karwar, **Maharashtra**: Mumbai, **Andaman and Nicobar Islands.**

Abroad: Circum-tropical.

Remark: This species is estuarine and intertidal can live within the salinity range of 20 ppt to 35 ppt embedded in sandy substratum. Usually seen in the outer channel area only during the low tide periods of summer seasons. Its availability in the lake proper is very less. Very less information is gathered about this species from Chilika lake. They crawl on the substratum in order to feed on settled planktonic materials. Dead shells are used in the decorative purposes and for the preparation of lime. In Chilika lake this is frequently observed in the high salinity period i.e. during the summer season at the outer channel area. Its distribution is restricted to outer channel area up to Satapada region only. It is never reported in the lake proper. Its muscle is used by the shore birds and other carnivorous fishes and shell fishes as food. Thus play a critical role in the coastal food chain.

Pugilina cochlidium

Phylum : Mollusca

Class: Gastropoda

Order : Neogastropoda

Family : Melongenidae

Genus and species : *Pugilina cochlidium* Linnaeus, 1758

Pugilina cochlidium

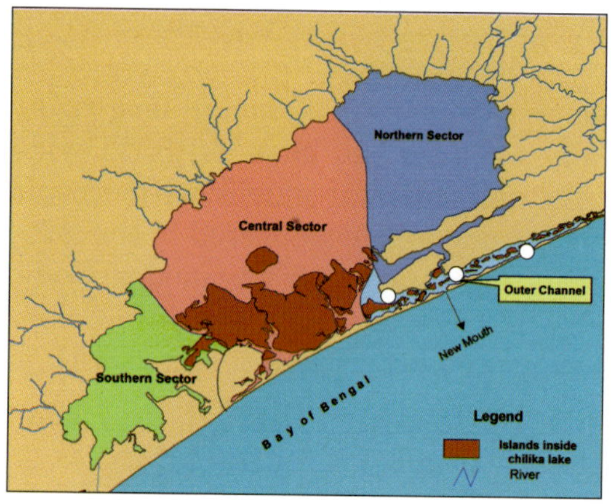

Distribution in Chilika lagoon

Description: Measurement: L = 40-88, D =20-51; Shell large, redish brown in colour, pear shaped; Spire high; Whorl angulate; aperture narrow and elongate. Presence of axial ribs on the spire whorls; presence of about eight compressed tublecles at the angular shoulder of the last two whorls.

Distribution in Chilika : In Chilika lake this marine gastropod is recorded from outer channel area only. Some dead shells are occupied by the hermit crab. Strictly seen during the summer seaon and widely seen during the lowtide region at outer channel area.

Distribution in India:. It is also recorded from **Orissa**: Subarnarekha Mouth; Talasari, Chandipur, Dhamra estuary; Paradwip, Mahanadi Estuary, Konark; Chilika lake, Ruashikulya estuary, Chhatrapur, **West Bengal**: Diamond Harbour; Sundarbans,; Digha Coast; Ganga Sagar, Andaman and Nicobar Islands, Gujarat, **Kerala**, **Maharashtra**, **Tamil Nadu**: Parangipettai Coast; Gulf of Mannar; Arukkattuthurai; Kodikkarai; Mallipattinam;

Manalmelkudi; Adirampattinam; Adyar Estuary; Vellar Estuary; Cooum Estuary; Pulicat Lake; Pamban; Porto Novo, Tellicherry.

Abroad: Red sea to Philipines and Japan.

Remark: This species is common representative all along the east coast and west coast of India. Dead shells are used as the decorative purposes as well as for the lime formation.

Stellaria solaris

Phylum : Mollusca

Class: Gastropoda

Order : Caenogastropoda

Family : Xenophoridae

Genus and species : *Stellaria solaris* Linnaeus, 1764

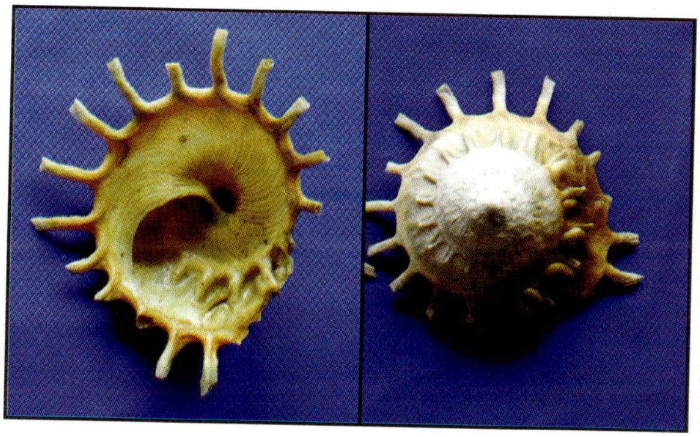

Stellaria solaris

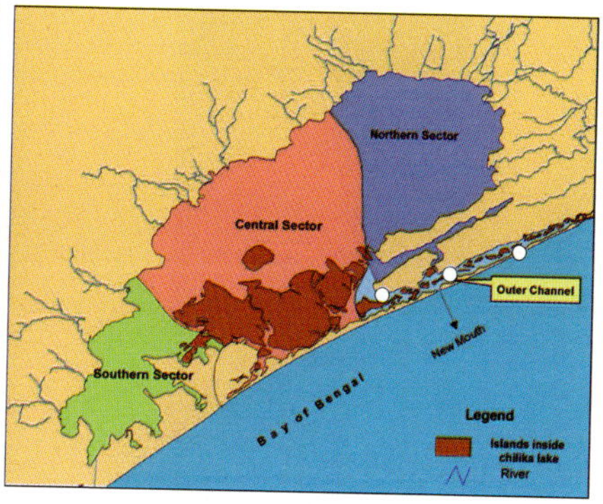

Distribution in Chilika lagoon

Description: Height 26.70-18.05 Width 59.40-43.70 Height of the aperture 36.20-26.00. Shell medium, orbicular, thin, and deeply conoid; spire low; whorls convex with hollow blunt protruding spines at the periphery obscuring the suture; aperture oval. Base almost flat; umbilicus deep, broad, with string wavy, uneven nodulose.

Body Colour: Cream, grey, light brown or shiny dark brown.

Distribution in India : Andhra Pradesh, Andaman Islands, Goa, Gujarat, Kerala, Maharashtra, Orissa, Tamil Nadu, West Bengal.

Abroad: Red Sea, Gulf of Aden, Gulf of Oman, Pakisthan Coast, Maldives, Myanmar Thailand, Malaysia, Indonesia, Philippines.

Remark: This species is uncommon to the Odisha coast as well as to Chilika lake. It is first time to record from the lagoon premises after the opening of the artificial lagoon inlet. Initially it is believed that *S. solaris* might be washed in to the lake due to the tidal incursion or wave action. But, frequent occurrence of live shells during the study period confirmed that this species is already accommodated in the lake ecosystem.

Telescopium telescopium

Phylum : Mollusca

Class: Gastropoda

Order : Caenogastropoda

Family : Potamididae

Genus and species : *Telescopium telescopium* Linnaeus, 1758

Telescopium telescopium

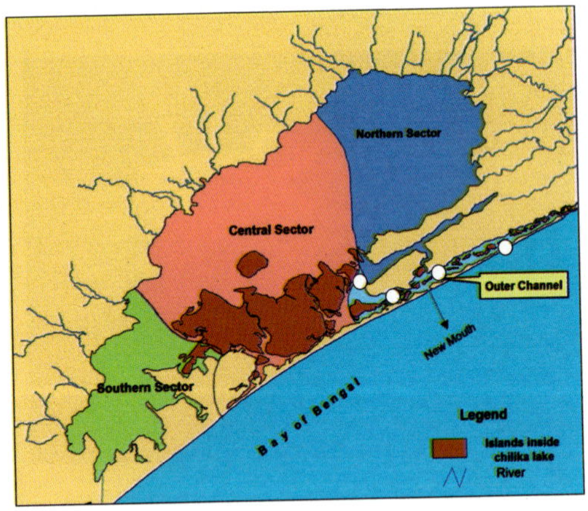

Distribution in Chilika lagoon

Description Size: 2 - 11 cm.Heavy, conical shell, twisted columella, shallow sutures, sides straight, base broad and rather flat, periphery angulate, whorls short, sculptured with several deep spiral striate, varixes lacking, anterior canal and columella very short, columella twisted, with a strong central spiral ridge and deep channel around the pillar, outer lip thin, not flared, smooth except for weak serrations anteriorly.

Body Colour: Dark brown, often with a single spiral central white line on each whorl, columella yellow, interior brown.

Distribution in Chilika lake: Old mouth (Arakhkuda), New mouth (Shipakuda), Satapada, Rambharatia Is.

Distribution in India:. It is also recorded from **Orissa**: Subarnarekha Mouth; Talasari, Chandipur, Dhamra estuary; Paradwip; Mahanadi Estuary; Konark; Chilika lake; Ruashikulya estuary; Chhatrapur, **West Bengal**: Diamond Harbour; Sundarbans,; Digha Coast; Ganga Sagar, **Andaman and Nicobar Islands**, **Gujarat**, **Kerala**,, **Maharashtra**, **Tamil Nadu**: Parangipettai Coast; Gulf of Mannar; Arukkattuthurai; Kodikkarai; Mallipattinam; Manalmelkudi; Adirampattinam; Adyar Estuary; Vellar Estuary; Cooum Estuary; Pulicat Lake; Pamban; Porto Novo, Tellicherry.

Abroad: Mayanmar; Indonesia; Malaya Peninsula; Madagaskar; North Australia; Philippines; Singapore and Sri Lanka.

Remark: This species is common representative all along the east coast and west coast of India. Dead shells are used as the decorative purposes as well as for the lime formation. Even in Chilika also live shells are occurred in the mangrove dominated areas of outer channel area.

Thias Bufo

Phylum : Mollusca

Class: Gastropoda

Order : Neogastropoda

Family : Muricidae

Genus and species : *Thias Bufo*

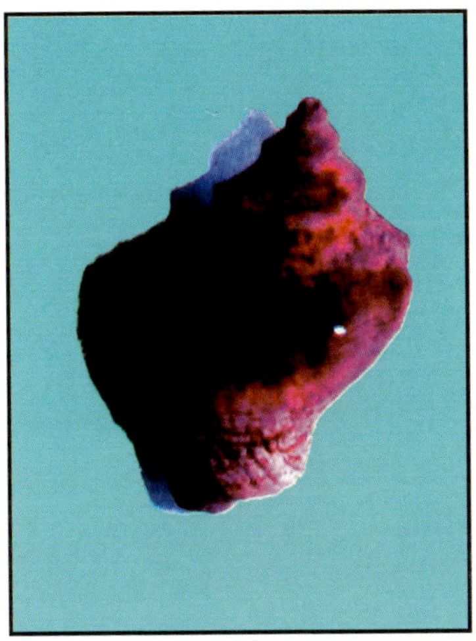

Thias Bufo

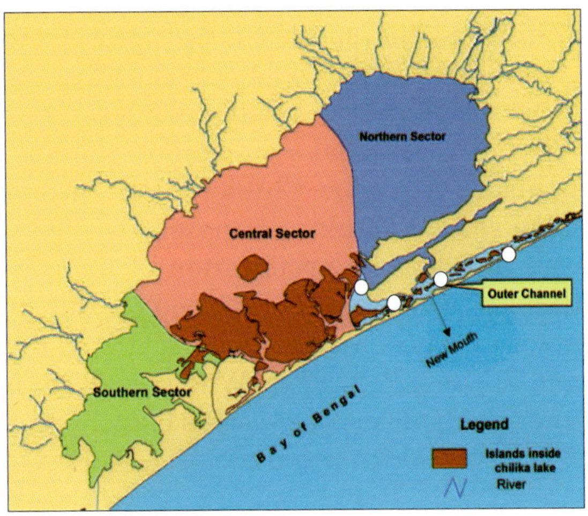

Distribution in Chilika lagoon

Description: The shell is thick, large and much broader with a very low spire and more or less broadly inflated body whorl. The surface of the shell is uniformly spirally grooved throughout and on the body whorl there are two spiral rows of rather widely set, rounded tubercles, those of the upper row, whose site is marked by an obtusely angular shoulder, being the larger; below these rows of nodules there are two slightly raised spiral ridges, but they are not tuberculated. The aperture is broad with a slightly arched columellar border covered by a thick, polished callosity which is particularly prominent posteriorly. The outer lip is thick, finely crenulated within and deeply notched at the upper extremity. The operculum is broadly ovate.

Body colour : Color: The shell is pale brown, faintly banded with dark brown on the surface and the columella and interior of the aperture are yellowish brown.

Distribution in Chilika lake : Old mouth (Arakhkuda), New mouth (Shipakuda), Satapada, Rambharatia Is.

Distribution in India : Odisha, Mahanadi Estuary; Chilika Lake, Gopalpur; Chhatrapur; Rushikulya estuary; Hukitola; Chandipur; **West Bengal, Andhra Pradesh, Gujarat** : Gulf of Kachchh,, **Kerala**,, **Tamil Nadu** : Ennore Backwaters, Gulf of Mannar; Krusadai; Shingle Island, Kerla, Kochi; Kozhikode, **Karnataka**: Karwar, **Maharashtra**: Mumbai, **Andaman and Nicobar Islands.**

Abroad: **Indo-West Pacific**: from East and South Africa to Papua New Guinea; north to Japan and south to Queensland.

Remark: This species is estuarine and intertidal can live within the salinity range of 20 ppt to 35 ppt embedded in sandy substratum. Usually seen in the outer channel area only during the low tide periods of summer seasons. Its availability in the lake proper is very less. Very less information is gathered about this species from Chilika lake. They crawl on the substratum in order to feed on settled planktonic materials. Dead shells are used in the decorative purposes and for the preparation of lime. In Chilika lake this is frequently observed in the high salinity period i.e. during the summer season at the outer channel area. Its distribution is restricted to Satapada region only. And never reported in the lake proper. When dead, its muscle feed on by the shore birds and other carnivorous gastropods even by the small fishes. Thus play a critical role in the coastal food chain.

Turritella attenuata

Phylum : Mollusca

Class: Gastropoda

Order : Caenogastropoda

Family : Turritellidae

Genus and species : *Turritella maculata* Reeve, 1849
Turritella columnaries Kiener, 1843
Turritella attenuata Reeve, 1849,
Turritella terebra Linnaeus, 1758

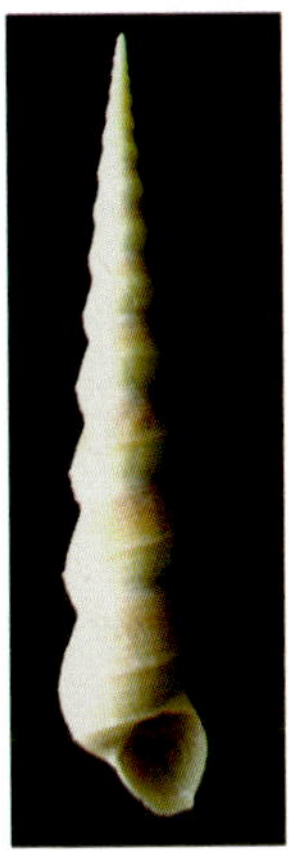

Turritella attenuata

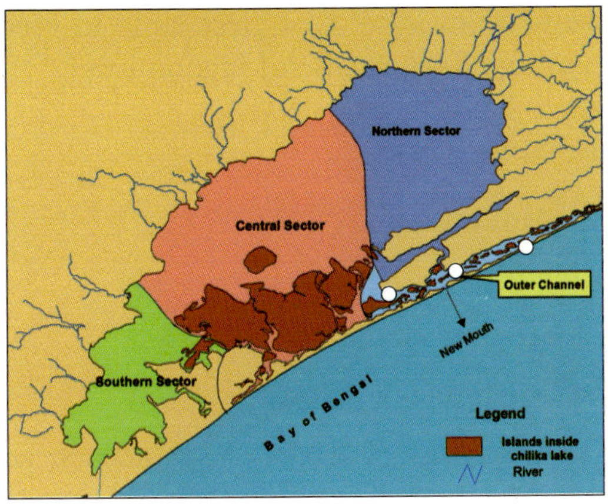

Distribution in Chilika lagoon

Description Size: Length: 35.75-83.00 mm,Diameter: 12-21.25 mm, Max height of the aperture: 15.32 mm. The shell is comparatively slender and considerably more elongated than in the preceding species, its height being at least four times as much as the maximum width of the body whorl. The whorls are more conspicuously bulged in the middle, and the sutural constrictions consequently appear deeper. The surface is traversed by fine spiral striae as well as by strongly raised spiral ridges, but the latter are, as a rule more numerous and are disposed differently; the strongest rib occurs somewhere near the middle and is followed above by a series of ribs which get successively weaker, and below by another raised rib nearly as strong as the medium one and separated from it by a rather wide interval.

Body Colour : The shell is browning, but more often tends to be more or less strongly tinged with blue.

Distribution in Chilika: In Chilika lake this marine gastropod is recorded from outer channel area only. Some dead shells are occupied by the hermit crab. Strictly seen during the summer seaon and widely seen during the lowtide region at outer channel area.

Distribution in IndiaOdisha, Mahanadi Estuary; Chilika Lake, Gopalpur; Chhatrapur; Rushikulya estuary; Hukitola; Chandipur; **West Bengal, Andhra Pradesh, Gujarat** : Gulf of Kachchh,, **Kerala**,, **Tamil Nadu** : Ennore Backwaters, Gulf of Mannar; Krusadai; Shingle Island, Kerla, Kochi; Kozhikode, **Karnataka**: Karwar, **Maharashtra**: Mumbai, **Andaman and Nicobar Islands.**

Remark: This is the first time ever to record this species from Chilika lake. It is a true marine species. This species is uncommon to the Odisha coast as well as to Chilika lake. It is first time to record from the lagoon premises after the opening of the artificial lagoon inlet.

Umbonium vestiarium

Phylum : Mollusca

Class: Gastropoda

Order : Vetigastropoda

Family : Trochidae

Genus and species: *Umbonium vestiarium* Linnaeus, 1758

Umbonium vestiarium

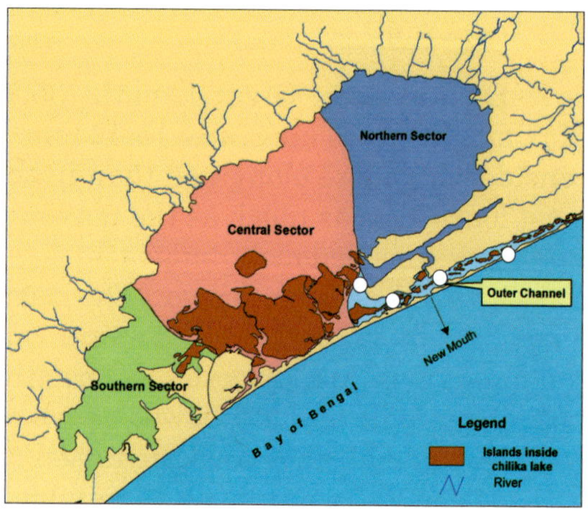

Distribution in Chilika lagoon

Description Size: 6-7mm high, 12-15 mm wide., Length: 4.7-8.4 mm, Diameter: 8.60-14.35 mm. Maximum height of the aperture: 3.40-4.30 mm. Solid, rounded shells, Small, disc-shape, flat-base, periphery rounded, whorls glossy smooth and lacking any sculpture, with a glossy, highly variable and colourfully patterned exterior. Conical shape, umbilicus closed by a thick callus, round body whorl, curved columella.The spire is depressed. The aperture is 'D' shaped. The umbilicus is filled with whitish callus which spreads over the basal surface of the shell.Polymorphism in shell color and pattern is known in this snail.

Body Colour: Grey, rose, yellow, pink, black, oliver-green or combination of color.

Distribution in Chilika : In Chilika lake this species is recorded from Magarmukh, new inlet region, Old inlet region (Arakhkuda).

Distribution in India: This species is also recorded from **Orissa**: Subarnarekha Mouth; Talasari, Kirtania; Chandipur; Dhamra estuary; Paradwip, Mahanadi Estuary, Konark; Chilika lake, Ruashikulya estuary, Chhatrapur, **West Bengal**: Diamond Harbour; Sundarbans,; Digha Coast; Ganga Sagar; Port Canning; Medinipur; North 24 Parganas; South 24 Parganas; Sankarpur; Old Digha; **Andaman and Nicobar Islands**, **Gujarat**, **Kerala**,, **Maharashtra**: Malvan; Ratnagiri; Mumbai; Mahim Creek; Alibag; Nariman Point; Kalbadevi Estuary; Bhatia Creek; Tarkarli Creek, Goa: Siridao beach, Siolim, Ribandar, **Karnataka**: North Kanara; Karwar; Kodibag; Ankola; Tadri, **Pondicherry**: Pondicherry mangrove; **Tamil Nadu**: Parangipettai Coast; Gulf of Mannar; Arukkattuthurai; Kodikkarai; Mallipattinam; Manalmelkudi; Adirampattinam; Adyar Estuary; Vellar Estuary; Cooum Estuary; Pulicat Lake; Pamban Kerala; Porto Novo, Tellicherry.

Abroad: Circum-tropical i.e. from Africa to Japan.

Remark: This species is typically constituent of the entire outer channel area. Often found attached in the sea grass or crawling of the substratum of the lake. Its flesh is eaten by the birds and the dead shell is used for the production of lime. Flesh is also used for the preparation of the food pellet for the aquaculture firms. Not edible by the locale people. During winter season i.e. from October to February higher abundance is noticed.

7

Summary and conclusion

Coastal lagoons, like estuaries are the transitional ecosystems between the land mass and sea. They retain many characteristic features of estuaries. Researchers had advocated much definition for a coastal lagoon. But the most acceptable definition is given by Kjerfve (1994). According to Kjerfve (1994) a coastal lagoon can be described as "*shallow coastal water body separated from the ocean by a barrier connected, at least, intermittently to the ocean by one or more restricted inlets.*" These are highly productive and sensitive ecosystems often stressed due to many natural and manmade activities (Alongi 1990). Since, they are located at the close vicinity of marine influence and influence of continental land mass, often typical brackish water condition prevails in them and thus become home to a large variety of flora and fauna of marine fresh water and brackish water forms. The low saline region (Oligohalaine) becomes favourable for migration of the anadromous fishes from sea in to the lagoons for breeding and spawning. These coastal ecosystems are generally rich in fisheries resources and provide huge socio-economic benefit to the state. Besides this, there are many other ecosystem services can be derived from coastal lagoon such as (i) protection from storms and floods (ii) flood control, (iii) reduces excess of air temperature by evaporative cooling, (iv) recharge ground water table, (v) major sink of nutrients and pollutants, (vi) functions as a sediment trap areas (vii) control of the sea level rise effects etc. In addition, the unique combination of fresh water, brackish water and marine conditions provides excellent homage to the top level trophic level representatives such as migratory birds and dolphins, sea cows etc.

Most of the coastal ecosystems are controlled by a number of physical, chemical, geological and biological factors which affects their ecology directly. Factors significantly affecting the coastal lagoons are their (i) geomorphology, (ii) plate tectonics, (iii) source inputs of sediment, (iv) fresh water influx, (iv) evaporation : precipitation ratio, (v) sea water intrusions (vi Pollution and other manmade effects, (vii) climate of the region like wind velocity& direction, Precipitation etc (viii) tidal incursion and coastal processes.

Coastal lagoons are unevenly and widely distributed all over the world from tropics to sub-polar regions and they occupy about 13 % of the total coast line. The Indian coast line is endowed with a large number of sensitive marine ecosystems in which coastal lagoons share a significant part. There are 17 coastal lagoons found along the coast lines of India in which eight coastal lagoons such as (1) Chilika lagoon, (2) Pulicat lagoon, (3) Pennar lagoon, (4) Bendi lagoon, (5) Nizampatnam lagoon, (6) Muttukadu lagoon, (7) Muthupet lagoon, (8) Gulf of Mannar lagoons are situated on the east coast, while the remaining nine lagoons such as (1) Vembanad lagoon, (2) Ashtamudi lagoon, (3) Paravur lagoon,(4) Ettikulam lagoon, (5) Veli lagoon, (6) Murukumpuzha lagoon (7) Talapady lagoon, (8) Lagoons of the Mumbai coast, (9) Lagoons of Lakshadweep are located on the west coast. Out of these 17 lagoons, the Chilika Lake is the largest in Asia and second largest in the World (First being Lake Baikal). Chilika Lake is immensely rich with biodiversity comprising fresh water, brackish water and marine forms.

The first scientific exploration of Chilika Lake was started during early period of twentieth century when Annandale, Kemp, Preston, Fauvel had started to document the baseline information on hydrography and biodiversity of the lake ecosystem. Perusal of literature especially the reviews made by Panigrahy (2000), Panda *et al,* (2008) and Sahu *et al,* (2014) shows that extensive studies relating to various aspects of the Chilika lagoon have been carried out over the last 100 years. The information referred in these reviews further delineates that much emphasis was laid to study the hydrography, nutrient dynamics, plankton & productivity, fisheries and nesting behaviour of migratory and resident birds and off late some studies have been made on environmental degradation including pollution of the lake water. The macrobenthos studies of this lake however, are limited only to preparation of the check list of some faunal groups.

The word 'Benthos' refers to the organisms, which lives on/ in/ in association with the bottom and water body. Based on their body size they are classified (Mare, 1942) in to three broad groups such as (i) Macrobenthos (> 500 μ) (ii) Meiobenthos (< 500 μ to > 63 μ) and (iii) Microbenthos (< 63 μ). Hence, macrobenthos are those organisms whose body size is > 500 μ. They are used as the best source of food for the birds, fishes and shellfishes and large bodied benthic invertebrates and thus, play a very crucial role transfer the energy from benthic habitants to large predators. Again, they play a very crucial role in the nutrient recirculation and maintain water clarity suspension feeding mechanisms. Due to their sessile mode of life style some researchers (Dauer, 1993) have suggested them as indicators of pollution. Gaufen and Tarzwell

(1952) opined that the indicator species must have (i) high numerical density, (ii) few species in a faunal assemblage, (iii) exhibit scavenger type of feeding and (iv) tolerance to low dissolved oxygen.

The pollution stress of the Chilika Lake is increasing day by day. The Major stressors are the organic pollution which mediated through the entry of untreated sewage causing nutrient enrichment, release of petroleum hydrocarbon due to the increased motorised boat operation for fishery and tourism, plastic litter falls and growth of noxious macrophytes especially, *Phragmites karka* infestation in northern sector that reduces the water transparency and generate enormous amount of organic debris to the sediment subsequently leading to anoxia. However the extent of pollution and their impact on biodiversity has not been studied yet. Since, macrobenthic organism reflects the health of the ecosystem and also provides basic information whether the system is severely polluted or moderately polluted or remains un-polluted. Hence this study was carried out. During the previous studies on (Patanaik, 1971; ZSI, 995) attempts were made only to develop species list of different macrobenthos occurring in the Lake.

Towards the end of the 20th century the ecological condition of the Chilika was severely degraded due to chocking of lagoon inlet. The salinity drop encouraged for the spreading of noxious weeds vegetation, migratory birds congregation had declined, the total capture fisheries catch reduced alarmingly and loss of biodiversity occurred. In order to revive the lake ecosystem from the edge of degradation, an artificial inlet was dredged on 23rd of September 2000 based on the recommendations of CWPRS, Pune. This had benefited the lagoon ecosystem considerably (Panigrahy 2006; Panigrahy *et al.,* 2009). The salinity level increased with fall in noxious weeds growth. Post hydrological intervention studies made by different researchers have shown substantial information regarding the increase in the capture fisheries output (Mohapatra *et al.,* 2007). But, a comprehensive study on post hydrological impact on benthic molluscan faunal diversity was not documented yet. Under this back drop, the present research was carried out in order to describe the benthic molluscan fauna which could be the first baseline information after hydrological intervention.

The present finding can be summarised as the following points:

1. During April 2007 to March 2010 a comprehensive study on molluscan fauna was made which can be observed from the table no. 5
2. The checklist of molluscan fauna described from Chilika lagoon showed a total of 147 molluscan species were identified belonging to 3 classes 20 order, 67 families.

3. Highest species were recorded among the class gastropod (82 sps.) belonging to 9 orders and 40 families followed by class bivalves (64 sps.) under 10 order and 26 families.
4. Least species contribution was seen by the polyplacophorans representing 1sps. namely *Acanthochitona mahensis* comprising to 1 order and 1 family only.
5. Outer channel area exhibited the highest diversity of bivalves and gastropods
6. Many of the new species are recorded during the present study referring that the outer channel area was rich in molluscan diversity.
7. Most of the newly occurring species are of typical in marine forms. It means that after the opening of the new lagoonal inlet during 23rd September 2000 and 1st August 2008 the lagoon became more supportive for the congregation of marine species.
8. Among bivalves, out of nine the order Veneroida is comprised of 41 species under 13 families & becomes the most dominant order that cover 66 % of the total bivalve species observed in Chilika observed during this study.
9. The families those are fall under the order Veneroida are Veneridae Cardiidae, Carditidae, Cultellidae, Cyrenidae, Donacidae, Glauconomidae, Mactridae, Psammobiidae, Semelidae, Tellinidae, Trapeziidae, Ungulinidae. The family Veneridae itself was comprised of 12 species alone contributed a major percentage to the total bulk of macrobenthic fauna.
10. A number of species of bivalves are found dominant in the sea grass/ sea weed bed of lagoon Chilika. These are mostly found from the outer channel area, southern sector and Central sectoer and Palur Channel connected point.
11. Some of the dominant species are those of *Meretrix meretrix* Linnaeus, 1758, *Meretrix casta* Gmelin, 1791, *Sunetta scripta* Linnaeus, 1758, *Theora opalina* Preston 1914, *Macoma birmanica* Philippi, *Tellina texturata* Sowerby in Reeve's 1867, *Silliqua aculata* Linnaeus, 1756, *Mactra mera* Reeve, 1854, *Tegillarca granosa* Linnaeus, 1758, *Brachidontes striatulus* Hanley, 1843, *Brachidontes undulates* Dunker, 1857, *Perna viridis* Linnaeus, 1758, *Solen kempi* Preston, *Crassostrea cuttackensis* Newton & Smith, 1912, *Saccostrea cucullata* Born 1778, *Placuna placenta* Linnaeus, 1758, *Hyotissa hyotis* Linnaeus, 1758, *Cuspidaria chilkaensis* Preston .

12. Out of these *Crassostrea cuttackensis, Saccostrea cucullata, Placuna placenta, Hyotissa hyotis, Sunetta scripta, Meretrix meretrix, Perna viridis* are exclusively found in the high salaine region where the salinity is above22-35 ppt.
13. Species like *Theora opalina, Tegillarca granosa, Brachidontes undulates, Brachidontes striatulus, Macoma birmanica* are exclusively of brackish water in nature.
14. One fresh water species namely *Lamellidens marginalis* have the ability to produce pearl, commonly seen in the fresh water dominated region of northern sector and central sector.
15. One wood boring bivalve *Martesia striata* is frequently occurring in the wooden boats, different shoots of mangroves and other wooden structures are found in the lake premises.
16. The windowpane oyster *Placuna placenta* was recorded from the sandy bottom of outer channel area where s species namely *Brachidontes striatulus, Perna viridis, Hyotissa hyotis, Crassostrea cuttackensis, Saccostrea cucullata,* are acting as potential biofoulers.
17. Gastropods dominate the lagoon bed as revaled during this study keep bivalves to the second position.
18. Most of them are comprised of the order Neogastropoda which includes 30 species belonging to 10 families measuring a share of 37 % of the total species of gastropod.
19. The second most prevailing group was observed in the form of Caenogastropod which contributes 23 % of the gastropod having 19 species grouped under 10 families. Both of these orders jointly contributed 49 species (30 from Neogastropod + 19 from Caenogastropod).
20. Overall contribution when calculated it brings out a voluminous share of 60 % of the total gastropod species recorded during this study.
21. The order Neogastropoda was predominant in the study area. Ten families namely Cancellaridae, Conidae, Fasciolariidae, Harpidae, Melongenidae, Muricidae, Nassariidae, Olividae, Pyramidellidae, Turridae are found under this order. Whereas eight families belonging to order Caenogastropoda such as Epitoniidae, Ficidae, Litiopidae, Potamididae, Scaliolidae, Thiaridae, Turritellidae, Babyloniidae were recorded second predominant group in the lake substratum of Chilika.
22. The observed dominant species of gastropods found in the study period, are those of *Babylonia spirata* Linnaeus 1758, *Bellamya bengalensis*

Lamarck 1822, *Bulia vittata* Linnaeus 1767, *Cerithideopsilla cingulata* Gmelin1791, *Cerithideopsis costata* da Costa, 1778, *Indothias lacera* Born 1778, *Littorina littorea* Linnaeus 1758, *Murex tribulus* Linnaeus 1758, *Natica tigrina* Röding 1798, *Nassarius stolatus* Gmelin 1791, *Neritina smithii* Wood 1828, *Oliva oliva* Linnaeus 1758, *Olivancillaria gibbosa* Born 1778, *Phalium areola* Linnaeus 1758, *Pila globosa* Swainson 1822, *Stenothyra blanfordiana* Nevill 1880, *Tarebia acula* Muller 1774, *Telescopium telescopium* Linnaeus, 1758, *Turritella aculate* Reeve 1849, *Umbonium vestiarium* Linnaeus 1758.

23. Bivalve species those are recorded for the first time in the Chilika lake are *Perna perna, Pitar erycina, Crassostrea gryphoides* conspicuous to mention.

24. Among gastropods the newly occurring species as recorded during the present study were those of *Babylonia spirata, Babylonia zeylanica, Ficus aculata, Ficus gracilis, Tibia delicatula, Stellaria solaris, Tonna dolium, Tonna sulcosa, Tonna allium, Cronia contracta, Chicoreus virginus, Chicoreus microphyllus, Fusinus colus, Harpa davidis, Turricula javana, Aplysia dactylomela, Armina maculata Bufonaria echinata etc.*

Conclusion

In concluding remark, it is necessary to mention here that after opening of the lagoon inlet during the year 2000 and 2008 in Chilika, the lagoon bed is appeared as one of the suitable habitat for the congregation of some of the species those are never reported earlier from the lake ecosystem. This further signifies that the outer channel area and salinity dominated areas are more diversified with malacofaunal community. However; a sincere management step is required to increase the habitat condition of the northern sector north-western region of the central sector.

References

1. Adam W., 1939. Cephalopoda, II, Revision des especes Indo-Malaises du genre Sepia Linnaeus 1758. III – Revision du genre Sepiella (Gary) Steenstrup, 1880. Siboga-Expeditie. Resultatats des expedition Zoologiques, Botaniques, Oceanographiques et enterprises aux Indes Neerfandaises Orientales en 1899-1900, 55b: pp. 35–122.
2. Allen G., E. Mandelli, and J. P. F. Zimmermann., 1981. Physics, geology, chemistry. Pp. 29-50 in P. Lasserre and H. Postma, editors. Coastal lagoon research, present and future: proceedings of a seminar. UNESCO Technical Papers in Marine Science 32. United Nations Educational, Scientific and Cultural Organization, Paris, France.
3. Alongi D.M., 1990. The ecology of tropical soft-bottom benthic ecosystems Oceanogr. Mar. Biol. Annu. Rev 28 (3): pp. 381-496
4. Annandale N. and Kemp S., 1915. Introduction to the fauna of Chilika lake . Mem.Indian Mus.,5: pp.1-20.
5. Annandale N.,1915. Introduction to the fauna of Chilika lake . Mem.Indian Mus.,5
6. Annandale, N. (ed). 1915-1924. The fauna of Chilika lake. Mem. Ind. Mus., 5 (1-13): pp. 1-1003.
7. Anthony A., J. Atwood, P. August, C. Byron, S. Cobb, C. Foster, C. Fry, A. Gold, K. Hagos, L. Heffner, D. Q. Kellogg, K. Lellis-Dibble, J. J. Opaluch, C. Oviatt, A. Pfeiffer-Herbert, N. Rohr, L. Smith, T. Smythe, J. Swift, and N. Vinhateiro., 2009. Coastal lagoons and climate change: ecological and social ramifications in U.S. Atlantic and Gulf coast ecosystems. Ecology and Society 14(1): 8. [online] URL: http: //www.ecologyandsociety.org/vol14/iss1/art8/.
8. Apte D., 1992. A unusual species of Turbinella pyrum (L). J. Bombay nat. Hist. Soc. 89: pp. 267.
9. Apte D., 1993. Marine Gastropoda of Bombay - A recent survey. J. Bombay nat. Hist. Soc. 90: pp. 537–539.
10. Apte D., 1997. Record of Homalocanthus secunda (Lamarck, 1822) from Okha in Gulf of Kutch. J. Bombay Nat. Hist. Soc. 95 (3): pp. 526–527.
11. Apte D., 1998. The book of Indian shells. Bombay Natural History Society, pp. 114.
12. Apte D., 2009. Opisthobranch fauna of Lakshadweep Islands, India with 52 new records to Lakshadweep and 40 new records from India: Part 1. J. Bombay Nat. Hist. Soc. 106 (2): pp. 162–175.

13. Apte D., Bhave, V., Parasharya, D., 2010. An annotated and illustrated checklist of the Opisthobranch fauna of Gulf of Kutch, Gujarat, India with 21 new records from Gujarat and 13 new records from India: Part 1. J. Bombay Nat. Hist. Soc. 107 (1): pp. 14–23.

14. Arularasan S., Kasinathan, R., 2007. Molluscan composition at Vellar estuary, Porto Novo coast.Zoo's Print J. 22 (1): pp. 2546.

15. Babu Philip, M., Appukuttan, K.K., 1995. A check list of gastropods landed at Santhikulangara-Neendakara area. Mar. Fish. Inf. Ser. T & E., No. 138: pp. 9–10.

16. Bates R.L., and Jackson, J.A.. (ed.), 1987. Glossary of geology, 3rd Ed. American Geological Institute, Alexandria, VA; pp. 788

17. Bird E. C. F., 1993. Submerging coasts: the effects of a rising sea level on coastal environments. Wiley, Chichester, UK.

18. Brusca R.C. & Brusca, G.J., 2003. Invertebrates (2 ed.). Sinauer Associates. pp. 702. ISBN 0-87893-097-3.

19. Chapman A.D., 2009. Numbers of Living Species in Australia and the World, 2nd edition. Australian Biological Resources Study, Canberra. Retrieved 12 January 2010. ISBN 978-0-642-56860-1 (printed); ISBN 978-0-642-56861-8 (online).

20. Cheriyan P.V., 1968. A collection of Molluscs from the Cochin Harbour area. Proc. Symp. on Mollusca, Marine Biological Association of India, Part 1. pp. 121–136,

21. Copeland B.J., 1967. Environmental characteristics of hypersaline lagoons. Publications of the Institute for Marine Science (University of Texas) 12: pp. 207-218.

22. Das A.K., Mitra, S.C., Mukhopadhyaya, S., 1977. Notes on the occurrence of Telescopium mauritsi Butot (Mesogastyropoda: Mollusca) from India. Newsl. Zool. Surv. India 3 (3): pp.129–130.

23. Das A.K., Mitra, S.C., Mukhopadhyaya, S., 1981. Studies on some molluscan collection by the "Golden Crown" from the Bay of Bengal with a note on camouflage habit of a gastropod, Xenophora pallidula (Reeve). Proc. Zool. Soc. Calcutta 32: pp. 79–87.

24. Day J.H., 1980. What is an estuary .S.Africa,J.Sci., pp.76:198.

25. Dey A., 2006. Contribution to the knowledge of Indian Marine molluscs (Part – IV) Family Tellindiae. Rec. Zool. Surv. India., Occ. Paper No 249: pp. 1–124, pls 38.

26. Dey A., 2008. Commercial and medicinal important molluscs of Sunderbans. Rec. Zool. Surv. India Occ. Paper No. 286: pp. 1–54.

27. Dey M., Jamadar, Y.A., Mitra, A., 2005. Distribution of intertidal malacofauna at Sagar Island.Rec. Zool. Surv. India 105 (1-2): pp. 25–35.

28. Eliot C., 1916. Fauna of Chilka Lake, Mollusca; Nudibranchiata. Ibid., S (1) : 377-379.

29. Emery K.O., and Stevenson R.E., 1957. Estuaries and Lagoon, Geol, Sec. America Mem., 67: pp. 673-750.

30. Gopinadha Pillai C.S., Appukuttan, K.K., 1980. Distribution of molluscs in and around the coral reefs of the southeastern coast of India. J. Bombay Nat. Hist. Soc. 77 (1): pp. 26–48.

31. Haszprunar G., 2001. "Mollusca (Molluscs)". Encyclopedia of Life Sciences. John Wiley & Sons, Ltd. doi:10.1038/npg.els.0001598

32. Holme N.A. and McIntyre A.D., 1971. Methods for the studies of marine benthos. I.B.P. Handbook No.16. Blackwell scientific publication, Oxford and Edinburg pp. 334.

33. Hornell J., Tomlin, J.R.K.B., 1951. Checklist of marine and fluvialite mollusca of Bombay and neighbourhood. Appendix In: Indian molluscs. Bombay Natural History Society, Bombay, pp. 83-97.

34. Joshi M.C., 1969. The marine Mollusca of the Konkan Coast. J. Shivaji Univ. (Sci.) 2 (4): pp. 47–54.

35. Jothinayagam J.T., 1987. Cephalopoda of the Madras Coast. Technical Monograph Zoological Survey of India, Kolkata, 15: pp. 1–85.

36. Jothinayagam J.T., 1987. Cephalopoda of the Madras Coast. Technical Monograph Zoological Surveyof India, Kolkata, 15: pp. 1–85.

37. Kasinathan R., Shanmugam, A., 1985. Molluscan fauna of Pitchavaram Mangroves, Tamil Nadu. Proc. Nat. Symp. Biol. Util. Cons. Mangroves (L.J. Bhosle (Ed.) pp. 438–443.

38. Kjerfve B., 1994, Coastal lagoons, In: Kjerfve B., (ed.), Coastal lagoon processes, Amsterdam: Elsevier, pp. 1-8.

39. Kohn A.J., 1978. The Conidae (Mollusca: Conidae) of India. J. Nat. Hist. 12: pp. 295–335.

40. Lankford R.R., 1977. Coastal lagoons of Mexico: their origin and classification. Estuarine process In: Wiley M., (ed.), Academic Press Inc., Volume II, pp. 182-215.

41. Little L.; Fowler, H.W.; Coulson, J.; Onions, C.T., eds. (1964). "Malacology". Shorter Oxford English Dictionary. Oxford University press.

42. Mahapatra A., 2001. Molluscan fauna of Godavari estuary: Estuarine Ecosystem Series Zoological Survey of India, Kolkata, 4: pp. 55–82,.

43. Mahapatra A., 2008. Molluscan fauna of Krishna estuary: Estuarine Ecosystem Series, Zoological Survey of India, Kolkata.5: 105–173,

44. Mahapatra, A., 2010. Mollusca of Vamsadhara and Nzagavali estuaries, Andhra Pradesh. Zoological Survey of India, Kolkata. Estuarine Ecosystem Series 6: 47–72.

45. Mahapatro D., Panigrahy, R.C., Panda,& Mishra R.K., 2016. Malacofaunal study of Chilika lagoon, East coast of India. S. Proc. Zool. Soc. https://doi.org/10.1007/s12595-016-0204-9

46. Mahapatro Debasish, 2017. Study of Macrobenthos of Chilika Lake –A coastal lagoon on east coast of India. Ph. D Thesis. Berhampur University, Berhampur-760007. pp. 256.

47. Mahapatro Debasish, Panigrahi, R.C. and Panda, S., 2013. Coastal Lagoon: Present Status and Future Challenges, International Journal of Marine Science, Vol.3, (23): pp 178-186 , doi: 10.5376/ijms.2013.03.0023.

48. Mahapatro Debasish, Panigrahy R.C., Panda S. and Mishra R.K., 2015. Checklist of intertidal benthic macrofauna of a brackish water coastal lagoon on east coast of India: The Chilika lake, International Journal of Marine Sciences. (5) 33, 1-13.

49. Mahapatro Debasish., 2006. Studies on the Macrobenthos off the Dhamra river mouth, Orissa Coast . M.Phil Thesis Submitted to Berhampur University.

50. Mare, Molly F.1942. A study of a marine benthic community with special reference to the microorganisms. Journal of Marine Biological Association UK. 25(03):pp. 517 - 554 DOI.

51. Menon M.G., Pillai, C.S.G., 1996. Marine biodiversity conservation and management. Central Marine Fisheries Research Institute, Spl. Publ., Cochin, p. 204.

52. Mookherjee H.P., 1985. Contributions to the molluscan fauna of India, Part III. Marine mollusks of the Coromandel coast, Palk strait and Gulf of Mannar-Gastropoda: Mesogastropoda (Pt.2).Rec. Zool. Surv. India, Occ. Paper No. 75, 1–93.

53. Mookherjee H.P., Barua, S., 1989. Molluscan fauna of Manauli Island in relation to environmental niche. Rec. Zool. Surv. India 85 (4): pp. 527–531.

54. More N.H. and Slinn, D.J. 1984. The physical hydrology of a lagoon system on the Pacific coast of Mexico. Estuarine, Coastal and Shelf Science 19: 413426.

55. Nagabhushanam A.K., Rao, G.C., 1972. An ecological survey of the marine Fauna of Minicoy Atoll (Laccadive Archipelago, Arabian Sea). Mitt. Zool. Mus., Berlin 48 (2): pp. 266–324.

56. Namboodiri P.N., Sivadas, P., 1979. Monation of molluscan assemblage at Kavaratti Atoll (Laccadive). Mahsagar Bull. Nat. Inst. Oceanogr. 12 (4): pp. 1239–1246.

57. Narayanan K.R., 1968. On the opisthobranchiate Fauna of the Gulf of Kutch. Proc. Symp. Molluscs,Mar. Biol. Assoc. India Part 1, pp. 188–213.

58. Narayanan K.R., 1969. On two Doridacean Nudibranchs (Mollusca: Gastropoda) from the Gulf ofKutch, new to the Indian Coast. J. Bombay Nat. Hist. Soc. 68 (1): pp. 280–281.

59. Narayanan K.R., 1972. The sacred chank of India. Seafood J. 4 (3): pp. 25–27.

60. Nicholas M., and Allen G., 1981. Sedimentary process in coastal lagoons, UNESCO technical paper in Marine Science, 33: pp. 27-80.

61. Nichols M. M., and J. D. Boon III., 1994. Chapter 7: Sediment transport processes in coastal lagoons. Pages 157-219 in B. Kjerfve, editor. Coastal lagoon processes. Elsevier, Amsterdam, The Netherlands.

62. Panda S., Bhatta K. S, Rath K. C., Misra, C. and Samal, R. N., 2008. The Atlas of Chilika, Chilika Development Authority, Bhubaneswar, India, pp.133.

63. Panigrahi S.N., Wikner J., Panigrahy R.C., Satapathy K.K. and Acharya B.C., 2009. Variability of nutrients and phytoplankton biomass in a shallow brackish water ecosystem (Chilka Lagoon), India. Limnology, 10: pp. 73-85.

64. Panigrahy R.C., 2000. The Chilka lake—a sensitive coastal ecosystem of Orissa, east coast of India. J Indian Ocean Stud., 7(2&3): pp.222–242.

65. Pati Sameer Kumar, 2007. Studies of the mollusks of Rushikulya estuary, east coast of India. M.Phil thesis submitted to Berhampur University.

66. Patil A.M., 1954. Study of marine fauna of the Karwar Coast and neighbouring Islands. J. BombayNat. Hist. Soc. 50: pp. 549–558.

67. Patil A.M., Gopalkrishnan, M., 1960. Molluscan shells washed on the sandy beach at Suratkal,South Kanara. Proc. Indian Science Congress 47 (3): pp. 468.

68. Phelger F.B., 1969. Some general Features of Coastal lagoons. In:Coastal Lagoon A symposium,Mexico,Edited by A.A.Castanares and F.B. helger, Universidad Nacional Autonoma, pp. 5-25.

69. Phleger F B., 198. A review of some general features of coastal lagoons. In:Coastal lagoon research, past, present, and future, UNESCO Technical Papers in Marine Science, 33: pp. 7-14.

70. Ponder Winston F. & Lindberg, David R., 2004. "Phylogeny of the Molluscs". World Congress of Malacology. Retrieved 2009-03-09.

71. Prabhakara Rao K., 1968. On a new genus and some new species of opisthobranchiate gastropods of the family Eubranchidae from the Gulf of Kutch. Proc. Symp. Molluscs, Mar. Biol. Assoc.India, , Part 1: pp. 51–60.

72. Pritchard, D. W., 1967. Observations of circulation in coastal plain estuaries, in Estuaries, , edited by G. H. Lauf, Amer. Assoc. Advanc. Sci., pp. 37-44.

73. Radhakrishna, Y., Ganapati, P.N., 1969. Fauna of Kakinada Bay. Bull. Nat. Inst. Sci. India 38: pp. 689–999.

74. Raghunathan, C., Sivaperuman, C., Ramakrishna, 2010. Diversity of littoral corals and their associated Molluscs and Echinoderms in Andaman Sea, South Andaman. In: Ramakrishna, Raghunathan, C., Sivaperuman, C. (Eds.), Recent trends in Biodiversity of Andaman and Nicobar Islands. Zoological Survey of India, Kolkata, pp: 249–273.

75. Rajagopal A. S. & Mookherjee, H. P., 1982. Contributions to the molluscan fauna of India~ Part II. Marine moJ!luscs oil the Coromandci Coast, Palk Bay and Gulf of MannarGastropoda : Mesogastropoda (Partim,), Rec. Zool. Surv. India, Calcutta. Occasiona1 paper No. 28: pp. 1- 53.

76. Rajagopal A.S., Subba Rao, N.V., 1974. On chitons from the Andaman and Nicobar Islands. J. Mar. Biol. Assoc. India 16 (2): pp. 398–411.

77. Rajagopal, A.S., Mookherjee, H.P., 1978. Contribution to the molluscan fauna of India Part I. Marine molluscs of Coromandel Coast, Palk Strait and Gulf of Mannar - Gastropoda: Archaeogastropoda.Rec. Zool. Surv. India, Occ. Paper 12: pp. 1–48.

78. Ramakrishna and Dey, A., 2003. Manual on the Identification of Schedule Molluscs from India: Published by the Director, Zool. Surv. India, Kolkata, pp. 1-40.

79. Ramakrishna Dey, A., Barua, S., Mukhopadhya, A., 2007. Fauna of Andhra Pradesh, Marine molluscs: Editor-Director, 2007. Fauna of Andhra Pradesh, State Fauna' Series, 5(Part-7)-Marine Molluscs: (Published by the Director, Zool. Surv. India, Kolkata) pp. 1-260

80. Ramakrishna Sreeraj, C.R., Raghunathan, C., Sivaperuman, C., Yogesh Kumar, J.S., Raghuraman, R., Immanuel, Titus, Rajan, P.T., 2010. Opisthobranchs of Andaman and Nicobar Islands. Zoological Survey of India, pp. 196

81. Rao D.V., 2010. Field Guide to Coral and Coral Associates of Andaman and Nicobar Islands. Zoological Survey of India, pp. 83.

82. Rao D.V., Sastry, D.R.K., 2007. Fauna of Button Island National Parks, South Andamans, Bay of Bengal. Rec. Zool. Surv. India, Occ. Paper 270, pp. 1–54.

83. Ray H.C., 1948a. On a collection of Mollusca from Coromandel coast of India. Rec. Indian Mus. 46: pp. 87–122.

84. Ray H.C., 1948b. Revision of Cypracea in the collection of Zoological Survey of India. Part 1. The families Triviidae, Eratoidae and Pediculariidae. Rec. Indian Mus. 46: pp. 183–213.

85. Ray H.C., 1952. Cowries (Mollusca: Gastropoda: Family Cypraeidae). J. Bombay Nat. Hist. Soc. 49 (2): pp. 663–669.

86. Ray H.C., 1954. Mitres of Indian waters. Mem. Indian Mus. 14: pp. 1–72.

87. Roy M., Dey, A., Banerjee, S., Nandi, N.C., 2008. Molluscan Macrobenthic diversity of Backwater Wetlands in West Bengal. Zoological Research in Human Welfare, Paper-I pp. 25–34.

88. Sahu B. K., Pati, P. and Panigrahy, R. C., 2014. Environmental conditions of Chilika Lake during pre and post hydrological intervention: an overview. Journal of Coastal Conservation (Springer Science & Business Med;Jun2014, Vol. 18 Issue 3: pp. 285.

89. Satyamurti S.T., 1952. The mollusca of the Krusadai Island (In the Gulf of Manaar). Bull. Madras Govt. Mus. New Ser. (Nat. Hist) 1 (2): pp. 1–267, 6.

90. Satyamurti S.T., 1956. The Mollusca of Krusadai Island (in the Gulf of Mannar) II. Scaphoda,Pelecypoda and Cephalopoda. Bull. Madra Govt. Mus. N.S. Nat. Hist. Sec. 7: pp.1–202.

91. Satyanarayana Rao K., Sundaram, K.S., 1972. Ecology of intertidal molluscs of Gulf of Mannar and Palk Bay. Proc. Indian Nat. Sci. Acad. 38B (5&6): pp. 462–474.

92. Sewell R. B. S. and Annandale N., 1922. Fauna of Chilka Lake : The hydrography and invertebrate fauna of Rhambha Bay in an abnormal year—Stiliger pica Annandale & Prashad,sp. nov., Mem. Indian Mus., 5: pp. 700-702.

93. Smith S.V. 1993. Stoichiometry of C:N:P fluxes in shallow-water marine ecosystems. In: Cole, J., Lowett, G., Findlay, S., (Eds.), Comparative Analyses of Ecosystems. Patterns, mechanisms and theories. New York: Springer-Verlag, pp. 259-286.

94. Sourina A. and M. Ricard 1976. Phytoplankton and its contribution to primary productivity in two coral reef areas of French Polynesia.l:..Mar. Biol. Ecol., 21: pp. 129-140.

95. Starmuehlner F., 1974. Beitrage zur Kenntnis der Mollusken-Fauna in Littoral von sudindien und Ceylon. J. Mar. Biol. Assoc. India 16 (1): pp. 49–82.

96. Subba Rao N.V., 1968. Report on a collection of wood boring molluscs from Mahanadi estuary, Orissa, India. Proc. Symp. Moll. Mar. Biol. Assoc. India, Part 1. Pp. 85–93,

97. Subba Rao N.V., 1970. On the collection of Strombidae (Mollusca: Gastropoda) from Bay of Bengal, Arabian sea and Western Indian ocean, with some new records, 1. Genus Strombus. J. Mar. Biol. Assoc. India 12 (1&2): pp. 109–124.

98. Subba Rao N.V., 1980. On the Conidae of Andaman and Nicobar Islands. Rec. Zool. Surv. India 77: pp. 39–50.

99. Subba Rao N.V., 2003. Indian Seashells (Part I), Polyplacophora and Gastropoda. Rec. Zool. Surv. India, Occ. Paper 192: pp. 1–416.

100. Subba Rao N.V., Dey, A., 1975. Studies on Indian Mitridae (Mollusca: Gastropoda: Stenoglossa). Newsl. Zool. Surv. India 1 (4): pp. 79–80.

101. Subba Rao N.V., Dey, A., 1984. Contribution to the knowledge of Indian marine molluscs. 1. Family Mitridae. Zool. Surv. India, Occ. Paper 61: pp. 1–48.

102. Subba Rao N.V., Dey, A., 1986. Contribution to the knowledge of Indian marine molluscs. 2. Family Donacidae. Zool. Surv. India, Occ. Paper 91: pp.1–30.

103. Subba Rao N.V., Dey, A., 2000. Catalogue of marine molluscs of Andaman and Nicobar Islands. Rec. Zool. Surv. India, Occ. Paper 187: pp. 1–323.

104. Subba Rao N.V., Dey, A., Barua, S., 1983. Studies on the malacofauna of Muriganga estuary, Sunderbans,West Bengal. Bull. Zool. Surv. India 5 (1): pp. 47–56.

105. Subba Rao N.V., Dey, A., Barua, S., 1992. Estuarine and marine molluscs of West Bengal: StateFauna Series 3, Fauna of West Bengal (Part-9). Zoological Survey of India, Kolkata, pp. 129–268.

106. Subba Rao N.V., Dey, A., Barua, S., 1993. Mollusca. Estuarine Ecosystem Series Part 2, Hugli-Matla Estuary. Zoological Survey of India, Calcutta, pp. 41–91.

107. Subba Rao N.V., Mookherjee, H.P., 1975. On a collection of Mollusca from the Mahanadi Estuary, Orissa. In: Natarajan, R. (Ed.), Recent Research in Estuarine Biology. Hindustan Publishing Corp, Delhi, pp. 165–176.

108. Subba Rao N.V., Sastry, D.R.K., 2005. Fauna of Marine National Park, Gulf of Kachchh, (Gujarat) an overview. Conservation Area Series Zoological Survey of India, Kolkata. 23: pp. 1–79,

109. Subba Rao N.V., Surya Rao, K.V., 1980. On a rare Nudibranch, Thordisa crosslandi Eliot (Mollusca: Dorididae) from the west coast of India. Bull. Zool. Surv. India 2 (2&3): 1- 219.

110. Subba Rao N.V., Surya Rao, K.V., 1991. Mollusca of Lakshadweep. State Fauna Series, 2: Fauna of Lakshadweep. Zoological Survey of India, Kolkata, pp. 273–362.

111. Subba Rao N.V., Surya Rao, K.V., 1991. Mollusca of Lakshadweep. State Fauna Series, 2: Faunaof Lakshadweep. Zoological Survey of India, Kolkata, pp. 273–362.

112. Subba Rao N.V., Surya Rao, K.V., 1993. Contribution to the knowledge of Indian marine molluscs.3. Family Muricidae. Zool. Surv. India, Occ. Paper 153 pp. 1–133.

113. Subba Rao N.V., Surya Rao, K.V., Maitra, S., 1991. Marine molluscs of Orissa. Zool. Surv. India:State Fauna series, 1: Fauna of Orissa (Part 3): pp. 1–175.

114. Subba Rao N.V., Surya Rao, K.V., Manna, R.N., 1995. Molluscs of Chilka Lake. Wetland Ecosystem Series, 1, Fauna of Chilka Lake, Zoological Survey of India, Kolkata, pp. 391–468.

115. Subba Rao N.V., Surya Rao, K.V., Mitra, S.C., 1987. Malacological notes on Sagar Island. Bull.Zool. Surv. India 8 (1-3), pp. 149–158.

116. Subrahmanyam, T.V., Karandikar, K.R., Murti, N.N., 1952a. Marine gastropoda of Bombay. J. Univ. Bombay 20: 21–34.

117. Subrahmanyam, T.V., Karandikar, K.R., Murti, N.N., 1952b. Marine gastropoda of Bombay, Part II. J. Univ. Bombay 21 (3): pp. 26–73.

118. Surya Rao K.V., Maitra, S., 1998. Fauna of Mahanadi Estuary: Mollusca. Estuarine Ecosystem Series, 3, Fauna of Mahanadi Estuary. Zoological Survey of India, Kolkata, pp. 161–197.

119. Surya Rao K.V., Maitra, S., Barua, S., Ramakrishna, 2004. Marine Molluscs of Gujarat (Part I : Polyplacophora, Gastropoda and Scaphopoda). State Fauna Series 8, Fauna of Gujarat. Zoological Survey of India, Kolkata, pp. 263–331.

120. Surya Rao K.V., Subba Rao, N.V., 1991. Molluscan fauna of Lakshadweep. Zool. Surv. India: State Fauna series, 2, Fauna of Lakshadweep, pp. 273–362.

121. Tikader B.K., 1964. Marine fauna of Deogad Coast (Ratnagiri District), Maharashtra, 1. Mollusca.J. Univ. Poona (Sci. Technol.) 28: pp. 43–48.

122. Tikader B.K., Daniel, A., Subba Rao, N.V., 1986. Sea shore animals of Andaman and Nicobar Islands. Zoological Survey of India, Kolkata, pp. 1-188.

123. Tikader B.K., Das, A.K., 1985. Glimpses of animal life of Andaman and Nicobar Islands. Zoological Survey of India, Kolkata, pp. 170.

124. Tripathy B. and Mukhopadhyay A.K., 2014. Marine Molluscan Diversity in India in Marine Faunal Diversity in India. Ed. Zoological Survey of India, Kolkata, West Bengal, India. DOI: 10.1016/B978-0-12-801948-1.00004-5

125. Troussellier Marc (Lead Author);Jean-Pierre Gattuso (Topic Editor) "Coastal lagoon". In: Encyclopedia of Earth. Eds. Cutler J. Cleveland (Washington, D.C.: Environmental Information Coalition, National Council for Science and the Environment). [First published in the Encyclopedia of Earth March 12, 2007) <http://www.eoearth.org/article/Coastal_lagoon>.

126. Venkataraman K., Jeyabaskaran, R., Raghuram, K.P., Alfred, J.R.B., 2004. Bibliography and Checklist of Corals and Coral Reef associated organisms of India. Rec. Zool. Surv. India, Paper 226: 1–468, Occ.

127. Venkataraman K., Raghunathan, C., Sreeraj, C.R., Raghuraman, R., 2012. Guide to the dangerous and venomous marine animals of India. Zoological Survey of India, Kolkata, pp.1- 104.

128. Venkitesan R., Mukherjee, A.K., 2011. New records of Comitas albicincta (Adams and Reeve, 1830) and Turritella bicingulata (Lamarck, 1822) (Mollusca: Gastropoda: Turritellidae) from India. Rec. Zool. Surv. India 111 (2): 95–97.

129. Virabhadra Rao K., Krishna Kumary, L., 1973. On a new species of Dendrodoris Ehrenberg from Goa: (Molluscs: Nudibranchiata). J. Mar. Biol. Assoc. India 15 (1): pp. 242–250.

130. Virabhadra Rao K., Krishna Kumary, L., 1974. On some aspects of taxonomy, structure and early development of the nudibranchiate gastropod, Discodoris fragilis (Alder and Hancock). J. Mar. Biol. Assoc. India 16 (3): pp 689–699.

131. Virabhadra Rao K., Sivadas, P., Krishna Kumary, L., 1974. On three rare Nudibranch Molluscs from Kavaratti Lagoon, Laccadive Islands. J. Mar. Biol. Assoc. India 16 (1): pp. 113–125.

132. Zimmerman J. T. F., 1981. The flushing of wellmixed tidal lagoons and its seasonal fluctuation.Pages 15-26 in P. Lasserre and H. Postma, editors. Coastal lagoon research, present and future:proceedings of a seminar. UNESCO Technical Papers in Marine Science 32. United Nations Educational, Scientific, and Cultural Organization, Paris, France.

Glossary

Albino - shell lacking normal pigmentation.

Alochthonous- material that has been imported into an ecosystem

Anterior - region situated near the head. In gastropods: in front.

Anterior canal - expansion looking like a groove or a tube and serving to protect the siphon in gastropod shells.

Apertural - position relative to the aperture of gastropod shells.

Aperture - opening in gastropod shells.

Apex - extremity of a gastropod shell opposite to the anterior region; part of the shell built in earlier life.

Apical - situated at or near the apex of a gastropod shell.

Arakhkuda – Old inlet which was closed during 2000.

Autochthonous- Native to the place where found; indigenous.

Axial - direction forming a plane with main shell axis in gastropods.

Badkuda- Island located at the south western region of Chilika

Barunakuda- An Is. located at the conjunction point of Central Chilika and Outer channel area.

Basal - position relative to shell base.

Base - part of the gastropod situated in opposition to the apex.

Beak - same as umbo.

Bhargabi - A river connected with Chilika at northern catchment area

Bhasra-Mundia- Hill located at the western flank of central Chilika

Bhavakundaleswar- A popular temple of lord Shiva located just in front of the lagoonal inlet. A popular tourist site..

Bhusandapur- It is a fish landing centre located on the north-western bank of Chilika Lake

Bivalve - molluscs that have, among other features, shell comprised of 2 halves, or valves.

Body whorl - most anterior whorl of the gastropod shell, last and largest whorl.

Bottom of the shell - same as base.

Brackish water – An ecosystem whose salinity fluctuates from fresh water (0 ppt to marine water (35 ppt) having the average ph of 12 to 16 is called as brackish water.

Byssal - position relative to byssus.

Byssal gap - gap or opening sometimes present on the ventral margin of bivalve shells for passage of byssus.

Byssus - bundle of fibers secreted by some bivalves attaching the animal to the bottom. Mussels, some arks, and pen shells are attached to the substrate by byssus.

Callus - thickening of the shell, secondary, smooth, sometimes glazed, usually secreted on the parietal region of the columella.

Cancellate - cross-barred sculpture. In bivalves, radial and concentric elements will cross to produce a cancellate sculpture.

Central sector- It is the central large portion of the lake basically Brackish water in nature.

Chadheiguhaa- A rocky hill located at the southern tip of Chilika.

Chadheihagaa- A rocky Is. located at the central Chilika.

Chilika –Large brackish water wetland ecosystem located within the Ganjam, Puri and Khordha districts of Odisha and on east it connected with bay of Bengal by couple of lagoons.

Chomata - marginal crenulations in Ostreidae and Gryphaeidae, occurring all around the inner side of valves or only near the hinge, composed of small tubercles or ridgelets on the right valve, and corresponding pits on the left valve.

Chondrophore - depression in spoon-like form housing the internal ligament of some bivalve shells.

Coastal Lagoon- A shallow coastal water body separated from the ocean by a barrier connected, at least, intermittently to the ocean by one or more restricted inlets.

Columella - column or pillar located on the centre of a gastropod shell.

Compressed - outline of bivalves which are flattened laterally.

Concentric - feature of sculptural elements curving about the umbo in bivalves.

Cord - element of gastropod shell sculpture, usually spirally oriented, thicker than line.

Cordlet - narrow cord, thicker than line.

Corrugated - appearance of surfaces forming wrinkles.

Crenulated - appearance of surfaces that are delicately notched or corrugated. Term usually applied to wrinkled shell margin or edge.

Crenulations - notches, or wrinkles which are small and delicate.

Daya- A river connected with chilika at northern catchment area

Denticles - features of sculpture elements looking like small teeth-like projections.Term usually applied to features seen on the internal part of the aperture.

Depressed - outline of low, pressed-down gastropod shells. OBS: Term usually applied to some top shells or baby ear.

Dhalabali: Located at outer channel area in which another mouth was opened during the monsoon period of 2012.

Dhanua - A river connected with Chilika at western catchment area

Dorsal - region opposite to the foot in gastropods.

Elongate - shell with length significantly larger than width.

Epifauna - Benthic invertebrates crawls over the sediment.

Equivalve - characteristic of bivalves that have the 2 valves or halves of same size.

Escutcheon - an area of the posterodorsal shell surface near a ligament that is differentiated by sculpture and frequently demarcated by a marginal ridge or furrow.

Excavated - appearance of a hollow, concave surface.

Fold - ridge spiralling on columella.

Foliated - characteristic of being leaf-like.

Foot - in gastropods, fleshy, sole-like, muscular part of body involved in locomotion.

Furrow - groove in longitudinal direction found on the dorsal region of, among other shells, cowries and Triviidae.

Fusiform - characteristic of being spindle-shaped.

Gabakunda: Another Island village of Chilika in front of which natural inlet was opened on 1st of August 2008 .

Ghantasila- Hill located at the southern most flank of Chilika

Glassy - surface resembling glass, vitreous, transparent.

Globose - same as globular.

Globular - shape resembling a sphere or a ball.

Grab sampler – an instrument used for the collection of sediment from aquatic ecosystems. It comes in variety of surface areas.

Granulated - surface covered with minute grains, pustules, or beads.

Growth lines - lines on shell surface indicative of alternating periods of growth and rest; sometimes corresponding to seasonal changes.

Hinge - region of the bivalve shell where the two valves are joined together, usually including interlocking teeth and the ligament.

Hinge teeth - projections that interlock on the inner side of the bivalve shell hinge helping to prevent the two

Honeymoon Is.- Island located at the south western region of Chilika

Horny - substance that is hardened and proteinaceous; present in or completely forming the gastropod operculum and shell periostracum.

Incised lines - features of shell sculpture represented by cuts or narrow grooves on the shell surface.

Indentation - cut or notch on shell edge or parietal region.

Indented - surface bearing an indentation.

Inequivalve - characteristic of having the two valves (halves) of different size.

Infauna- Benthic invertebrates live inside the sediment or buried in form.

Inflated - characteristic of being 'fat', rotund, and frequently lightweight.

Interspaces - spaces between sculptural features, e.g., ribs, costae, or cords.

Juvenile - characteristic of being young, immature, not fully grown.

Kalijai- Island located at the central region of Chilika

Kalijugeswar- Hill located at the western flank of central Chilika . A shiv temple is alos present.

Kalupadaghat- It is a fish landing centre located on the north-western bank of Chilika lake

Kani- A river connected with chilika at western catchment area

Kankadakuda- An Is. located at the conjunction point of Central Chilika and Outer channel area.

Keyhole - apical orifice in some limpets.

Knob - large nodule, rounded projection.

Knobbed - surface bearing knobs.

Krushnaprasad Gargh- Island located at the eastern region of Chilika

Lake - A lake (from Latin lacus) is a large body of water (larger and deeper than a pond) within a body of landmass. Lakes do not flow, like rivers, but many have rivers flowing into and out of them.

Lamella - thin plate or blade-like projection

Lamellation - same as lamellae.

Ligament - structure that is horny and proteinaceous, acting as a spring tending to keep the valves opened in bivalve shells. Usually situated in the region of the hinge, either internally or externally.

Line - sculptural feature narrowly incised on shell surface.

Lip - edges of the outer surface of the aperture in the gastropod shell.

Longitudinal - direction parallel to the largest dimension of the shell or mollusc.

Luna- A river connected with chilika at western catchment area

Lunule - impression on the external side of the hinge, anterior to the umbo, usually heart-shaped.

Macrobenthic invertebrates- lower groups of animal without vertebral column ranged from the phylum Protozoa to Echinodermata.

Macrobenthos: benthic invertebrates those cannot penetrate 5oo micron of sieve size.

Makaraa- A River connected with Chilika at northern catchment area

Malaguni- A river connected with Chilika at western catchment area

Malatikuda- Small Island located at south of the Kalijai Is.

Mantle - fleshy sheet surrounding vital organs and composed of 2 lobes, one lining and secreting each valve.

Margin - edge of shell.

Meiobenthos : benthic invertebrates those can penetrate 5oo micron of sieve size but retained in 63 micron .

Microbenthos : benthic invertebrates those can penetrate 63 micron of sieve size

Nacreous - characteristic of being iridiscent, like mother-of-pearl.

Nalabana birds sanctuary- Muddy Island located at the central region of Chilika. Well known for the migratory bird congregation, feeding and nesting site.

NIO- National Institute of Oceanography

Nodules - projections that are rounded as tubercules.

Nodulose - surface bearing nodules.

Northern sector – Fresh water region of Chilika lake located at north western region.

Notch - cut or depression on any margin, canal, or on the gastropod aperture.

Opalescent - characteristic of being whitish, but with nacreous luster.

Operculum - trapdoor or plate which closes the aperture of gastropod shells and isolates the snail from its surrounding environment. Opercula can be horny ('soft', brownish) or calcareous ('hard', usually whitish).

Outer channel area- It is located on the eastern side of the lake connected with bay of Bengal by the couple of inlets. Salinity almost similar to bay of Bengal (i,e. 35 ppt.)

Outer lip - edge of the external part of the aperture away from the shell axis.

Oval/Ovate - characteristic of having the form of an egg.

Pallial line - fine scar-like impression present internally;in bivalve shells produced by the edge of the mantle.

Palur channel- A 16 kilometre long channel that joins the river Rushikulya at south with the southernmost tip of the Chilika lagoon.

Parietal - region of the internal part of the aperture, usually set apart by differences in surface texture and/or coloration.

Parietal shield - parietal region when markedly different from the remainder of the adjacent shell area.

Parikuda- Island located at the eastern region of Chilika

Periostracum - layer of the outside part of the shell. It is horny and sometimes hair-like.

Periphery - region of the outermost part of any given whorl on the gastropod shell. The shell periphery is therefore the greatest circumference of the gastropod shell.

Peristome - aperture rim or periphery.

Phragmites karka- A monocotyledon plant becomes menace to the Chilika lake ecosystem. It is an invasive reed found mostly in the northern and western flank of Chilika lake.

Plication - same as fold.

Posterior - in bivalves, the region of the shell sinus away from the foot.

Posterior - region away from the siphonal canal, near the apex, in gastropods; in bivalves, the region of the shell sinus, away from the foot.

Posterior canal - canal of small size or notch opposite to the siphonal canal on the aperture of the gastropod shell.

Prodissoconch - shell in larval state remaining on the umbonal region of well-preserved bivalve shells.

Protoconch - larval shell remaining on the apex of well-preserved gastropod shells.

Radial - structures that are directed away from the apex toward the shell margin, in limpets.

Rambha bay- Southern province of Chilika belonged to Ganjam and Puri districts.

Rambharatia- Sand bar located at close vicinity of artificial inlet. Infested with *Casuarina* plantation. A popular tourist site.

Ramsar – Place of Iran that formulates the rules and regulations related to wetlands.

Ratnachira- A river connected with chilika at western catchment area

Reticulate - feature of shell sculpture consisting of criss-crossed, net-like texture formed by the intersection of lines at right angles.

Reticulated - same as reticulate.

Ribbon - surface consisting of an aligned sequence of egg-cases.

Riblets - diminutive of ribs.

Ribs - structural elements forming a well-defined, narrow ridge in gastropod shells. Term usually applied to those elements forming a plane with (or slightly oblique to) shell axis.

Salia - A river connected with Chilika at western catchment area

Sanakuda- Island located at the south western region of Chilika

Satapada- Island inside the Chilika . Popular tourist site for Irrawaddy Dolphin

Scales - sculptural elements that are small, raised, and plate-like.

Sea grass- groups of monocot and dicot pants grows in saline water conditions

Sea weed- Lower groups of marine plants.

Septum - partition found in the internal side of gastropod shells; characteristic of slipper-shells.

Serrated - outline resembling tiny saw teeth.

Serrated - resembling tiny saw teeth.

Shell sinus - embayment on the pallial line of bivalve shells that correspond to the position of the siphons.

Shipakuda- Island village of Chilika in front of which the new lagoon inlet was made artificially during 23rd September 2000.

Shoulder - angled region of the whorls of gastropod shells.

Siphon - prolongation of the gastropod mantle used to convey water into the mantle cavity.

Siphonal canal - projection of the anterior region shell in tubular form protecting the anterior siphon.

Snails - common name of gastropods.

Somolonasi- Island located at the south western region of Chilika

Sonkhala lagoon- A brackish water lake located on Thailand.

Soran- It is a fish landing centre located on the western bank of Chilika Lake

Southern sector – Southern most province of the lake mostly saline in nature.

Spaced; evidence of a growth halt during which a thickened lip develops.

Spiral - direction following the coiling of the gastropod shell. Term usually applied as a modifier to sculptural terms such as 'spiral cords'.

Spire - series of successive whorls in a gastropod shell, with exception of the last one.

Spire angle - angle formed by the lines defined by the outermost points on both sides of the spire.

Striation - fine, repeated lines or furrows on shell surface.

Suture - line or region of junction between two adjacent whorls in the gastropod shell.

Synonym - a scientific name applied to a species that has received an earlier name. OBS: Usually, the earlier name is the valid one.

Thread - same as line.

Tinimuhani- Confluence point of river and Chilika Lake

Top of the shell - same as apex.

Trigonal - same as triangular.

Turbinate - form that looks top-shaped, tapering evenly from base to apex.

Turreted - form that looks tower-shaped, elongate.

Umbilicus - cavity at base of gastropod shells.

Umbo (pl. umbones) - projected portion of the hinge. OBS: First-formed part of the bivalve shell.

Uncoiled - gastropod shell that lacks coiling.

Valve - one half of the bivalve shell.

valves from sliding sideways past each other.

Varix - axial sculptural element that is more prominent than a costa, and usually more widely spaced; evidence of a growth halt during which a thickened lip develops (plural: varices).

Ventral - region of the animal opposed to the dorsal region; region of the foot in gastropods.

Ventral - region of the animal opposite the dorsal region; usually region of the foot in bivalves

Wetlands- A wetland is a land area that is saturated with water, either permanently or seasonally, such that it takes on the characteristics of a distinct ecosystem.

Whorl - a complete turn or coil of the gastropod shell.

Miscellaneous

Fresh water weeds of northern sector of Chilika lake (left image) and Phragmites karka at northern sector (right image)

Islands of Chilika lake Bhasra Mundia (left image) and Beacon Is land at southern sector (right image)

Image of artificially dredged inlets (upper two image) and Naturally opened inlet of Chilika lagoon (lower image)

Artificially implanted mangrove vegetation located at outer channel area of Chilika lake

Submerged sea weed vegetation of at Nalabana Bird Is. of Chilika lake

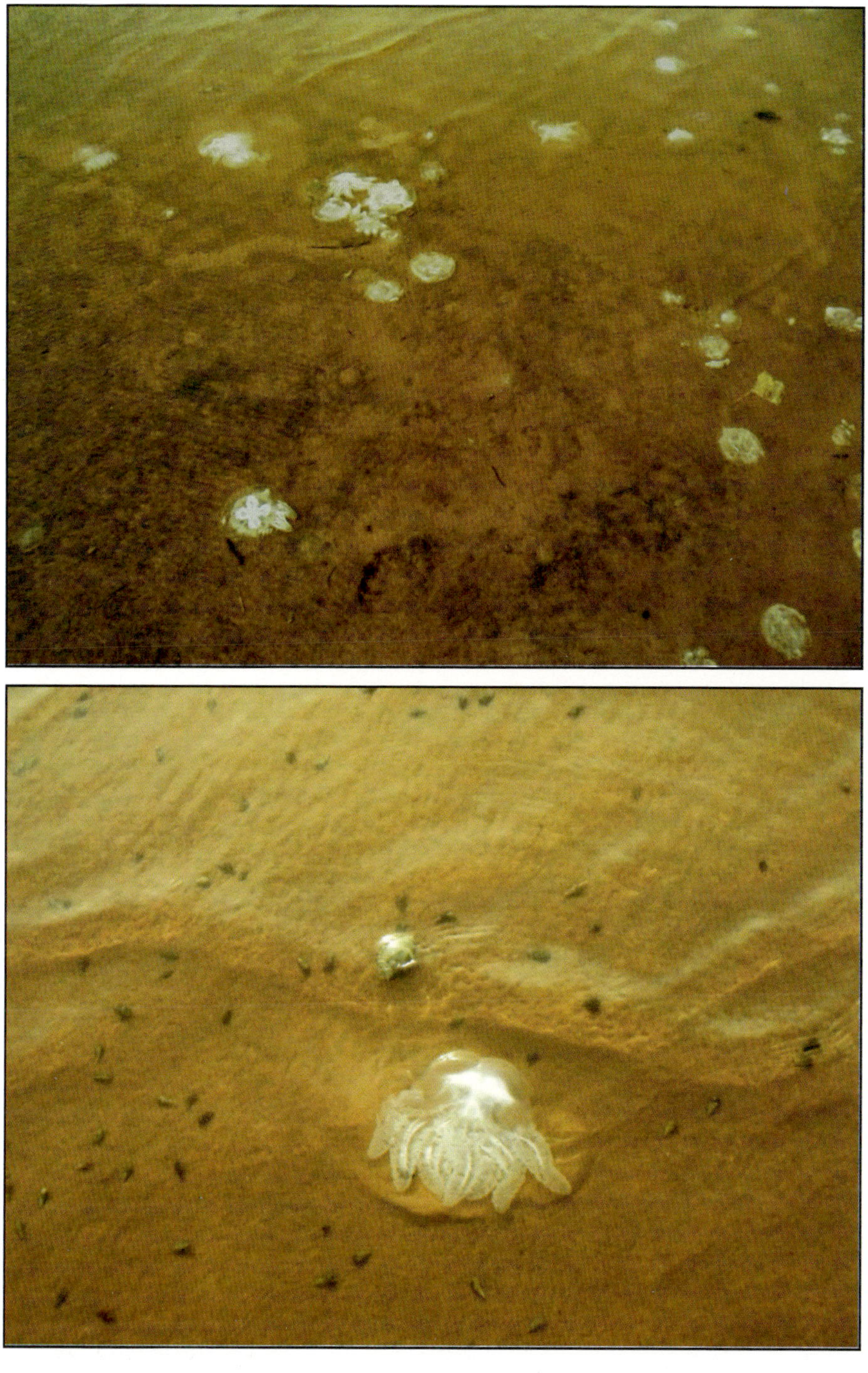

Jelly fish bloom in the outer channel area of Chilika lake

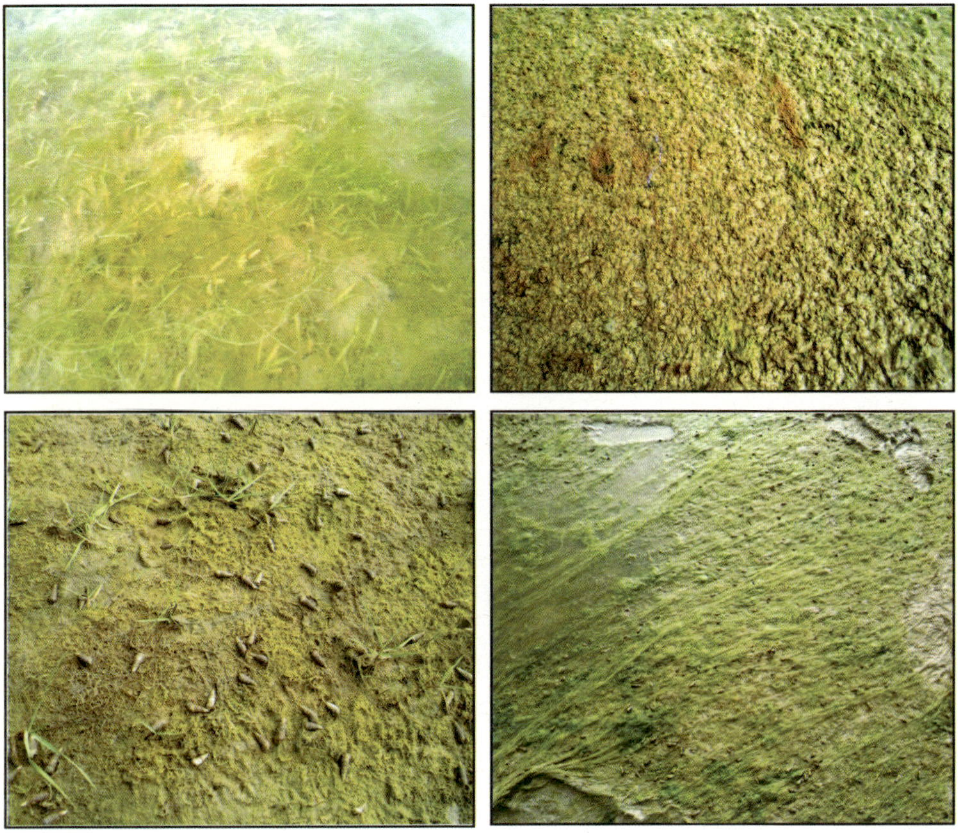

Sea grass and sea weed vegetation of Chilika lake

Sea grass Halophila ovalis and Halodule sp. at Southern sector of Chilika Lake

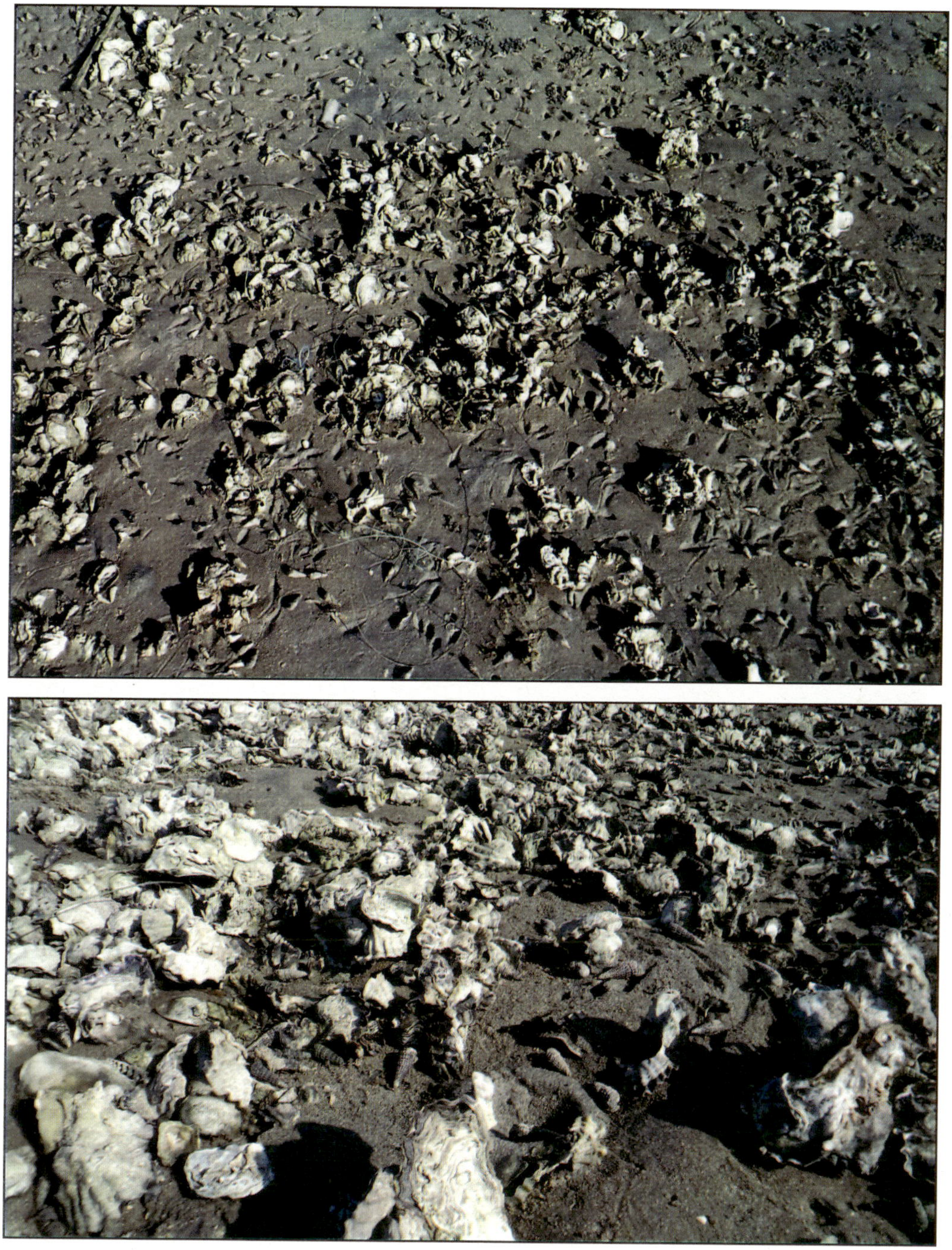

Mixed molluscan community in in a intertidal sand flat

Chilika lagoon inlet : upper image- Shipakuda inlet and lower image Gabakunda inlet

Intertidal mollusc at Gabakunda region

Intertidal mollusc at Arakhkuda region

Intertidal gastropod community upper image and bivalve community lower image at Bhabkundaleswar region of Chilika

Live Oyster shells collected from the intertidal region of Chilika lake

Gastropod burrowing in the sands of outer channel area

Molluscan shells driven by the waves and tides at outer channel area of Chilika lake.

Intertidal gastropod in a tidal pool of outer channel area of Chilika lake

First author standing at the Chilika inlet and Bay of Bengal confluence point